HOW TO
TAKE
OVER THE
WORLD

HOW TO TAKE OVER THE WORLD

PRACTICAL SCHEMES AND SCIENTIFIC SOLUTIONS FOR THE ASPIRING SUPERVILLAIN

RYAN NORTH

ILLUSTRATED BY CARLY MONARDO

Riverhead Books
New York
2022

RIVERHEAD BOOKS
An imprint of Penguin Random House LLC
penguinrandomhouse.com

Library of Congress Cataloging-in-Publication Data
Names: North, Ryan, 1980- author. | Monardo, Carly, 1984- illustrator.
Title: How to take over the world : practical schemes and scientific
    solutions for the aspiring supervillain / Ryan North ; illustrated by Carly Monardo.
Description: New York : Riverhead Books, [2022]
Identifiers: LCCN 2021048518 | ISBN 9780593192016 (hardcover) | ISBN 9780593192030 (ebook)
Subjects: LCSH: Science. | Supervillains. | Comic books, strips, etc.—Humor.
Classification: LCC Q172 .N67 2022 | DDC 502/.07—dc23/eng/20211110
LC record available at https://lccn.loc.gov/2021048518

International edition ISBN: 9780593541531

Printed in the United States of America
1st Printing

BOOK DESIGN BY LUCIA BERNARD

FOR LEX, VICTOR, ERIK, AND DR. ISLEY

Anyone who either cannot lead the common life or is so self-sufficient as to not need to, and therefore does not partake of society, is either a beast or a god.

—ARISTOTLE, POLITICS (350 BCE)

CONTENTS

PART THREE: THE UNPUNISHED CRIME IS NEVER REGRETTED

Oft have I digg'd up dead men from their graves,

And set them upright at their dear friends' doors,

Even when their sorrows almost were forgot;

And on their skins, as on the bark of trees,

Have with my knife carved in Roman letters,

"Let not your sorrow die, though I am dead."

Tut, I have done a thousand dreadful things

As willingly as one would kill a fly,

And nothing grieves me heartily indeed

But that I cannot do ten thousand more.

—FAMED PLAYWRIGHT WILLIAM SHAKESPEARE

(Okay, fine: the character of Aaron in Titus Andronicus,
***written** by famed playwright William Shakespeare)*

DISCLAIMER

This is a book about the edges of science, the limits of what's currently possible thanks to the technology that humans have already invented or are currently inventing, and the open questions that, once answered, will turn what's left of the impossible into the possible. It identifies hitherto unexploited weaknesses in our global civilization: cultural, historical, and technological blind spots that leave humanity susceptible to the maneuvering of a sufficiently motivated individual, one able to seize these opportunities to reshape the very fate of humanity itself.

In other words, this is a book of nonfiction about becoming a *literal supervillain* and *taking over the world*.

Normally here I'd add a "don't try this at home," but just to be safe: don't try this *anywhere*.

INTRODUCTION

HELLO AND THANK YOU FOR READING MY BOOK ABOUT WORLD DOMINATION

> Being brilliant's not enough . . . you have to work hard. Intelligence is not a privilege, it's a gift, and you use it for the good of mankind.
>
> —*Dr. Otto Octavius, a.k.a. "Doctor Octopus," in* Spider-Man 2 *(2004)*

A supervillain is normally considered the bad guy. I should know, I've written enough of them.

A big part of my work at companies like Marvel and DC Comics is to come up with new schemes for the villains. Whether I'm writing a wild and capricious ancient god, an undead wizard, a malevolent alien, a sinister warlord, a genius billionaire, a hunter of humans, or the greatest supervillain of them all (Doctor Doom),* they all need to reach for new and ever-more-audacious heights of supervillainy each and every month. And in order to sell the stakes of their stories to the reader, these schemes need to

---

*He's a time traveler, a scientist, *and* a magician. Even Lex Luthor, as great as he is, can't cast magic spells.

be credible. More than credible, actually: these heists have to *really work* . . . right up to the moment when they suddenly don't.

The truth of superhero comics—and this is common knowledge—is that no matter how hard the villains try, no matter how perilously close their schemes come to working, they will *always* be foiled at the last second, when the heroes dig deep and find some previously unknown well of strength or cleverness or compassion or robot suits. This follows the only universal law of storytelling: the closer the heroes come to losing— the more it seems that maybe this time the villain really will win—the more satisfying the heroes' victory will be. This law is so powerful that it applies to reality as much as it does to fiction: a lopsided hockey game is dull, but a hockey final where your team comes from behind in the last period and wins with a single dramatic shot in double overtime is anything but.

Like I said: common knowledge.

But what's *un*common knowledge is that if you take that universal law of storytelling and combine it with the fact that Marvel and DC are owned by the Walt Disney Company and AT&T's Warner Media LLC, respectively, then you uncover a terrifying truth. Two of the most powerful multinational corporations on the planet have spent decades, in plain sight, paying some of the most creative people alive today to design increasingly credible world-domination schemes—and these schemes have been thwarted only by *chance*, by *circumstance*, by the pins we writers have carefully inserted into our own grenades. Once you understand that, it doesn't take much to wonder: What if the supervillain didn't *have* to lose? What if the heroes faced a scheme so clever and bold and audacious and unprecedented that it could never be predicted, much less foiled? And if a supervillain could do that in fiction, what's to stop someone *else* from doing it in real life? That was the origin of this book: that moment when I realized that not only have I spent years working on ways to take over the world, but thanks to my background in science* . . .

. . . I've actually figured it out.

And unlike most supervillains, I'm more than happy to share.

WAIT, ARE THERE ANY ETHICAL CONCERNS WITH TELLING PEOPLE HOW TO TAKE OVER THE WORLD? IT'S JUST THIS SEEMS LIKE THE SORT OF THING THAT ETHICISTS MIGHT DESCRIBE AS "REALLY, *REALLY* WRONG."

This is a reasonable question, so let's get it out of the way early. Is it truly "unethical" (and "potentially dangerous") to provide detailed instructions for breaking the letter and spirit of local, national, and international laws, enabling readers to take the fate of the planet and everyone living on it into their own hands?

I mean . . . *maybe?* Expensive, highly trained lawyers and ethicists have said yes, but I already wrote this book and I say no way, so unfortunately, it's impossible to say where the truth lies.

Thankfully, we can sidestep all these tired arguments about "right" and "wrong" and "serious legal responsibility on the part of the author" by invoking the Precept of Hypotheticality! So here it is: I am explaining all this as a mere intellectual exercise intended to explore the limits of humanity's technology, history, and invention! Every scheme presented herein is entirely hypothetical, and I'm simply telling you how to take over the world *in your mind.* (Except for the one about cloning dinosaurs: that

---

*Before I began my superhero comics career, the University of Toronto accredited me as a Master of Science, capital "S" and everything. (Yes, my degree's in computational linguistics, but that also trained me to be *really good* at reading research papers.)

one's objectively awesome, despite what the frankly alarmist Jurassic Park franchise would have you believe, and you should do it at your earliest convenience.)

So relax! Enjoy the book! We're all here to have a good time, and if some of us end up gaining (in our heads) power over the fates of (imaginary) nations in the process, that's just the way things go sometimes.

SO LET'S TAKE OVER THE WORLD ALREADY.

You know what? *Let's.*

Our first advantage is this: nobody expects comic-book supervillain schemes to actually happen here in the real world. This makes it much less likely that any so-called "hero" will be ready, willing, or able to pop up at the last second to ruin these plots. And while here in reality we don't have access to incredible fantasy science like mind-control helmets and shrink rays, that's actually our second advantage: nobody *else* has those things either. A world without superheroes is actually more amenable to super-villainy than one crawling with them. In the film *Avengers: Infinity War*, before Thanos could achieve ultimate victory, he had to confront several superpowered fools standing in his way—including but not limited to a super-soldier, a rage monster, a billionaire playboy philanthropist in a flying iron suit, and a teenager with the proportional strength of a spider who shot webs out of his wrists.

We won't have to deal with any of that baloney.

But I may be getting ahead of myself with all this talk of "fools" and "ultimate victory." Let's first ensure we're all on the same page by examining what a supervillain actually *is*. The word seems simple enough: "super" means better, and "villain" means bad guy. So a supervillain is a better bad guy. End of story, right?

But then we look at Superman.

Superman can fly and leap tall buildings in a single bound (a lesser accomplishment than flying, but he always seems to mention it anyway). He's more powerful than a locomotive (which means he's got access to the power of about 12,000 horses, if we compare against some of the world's largest and most powerful locomotives); he's faster than a speeding bullet (which means he can move at least 120 meters per second [m/s], and if we consider high-velocity cartridges, that number jumps to 1,700 m/s); plus he can shoot lasers out of his eyes and freeze things with his breath. In terms of power and abilities, he's as far above the average person as we are above an ant. Clearly, "super" means more than just better here: it means going beyond. He's not just a stronger version of us: Superman can do things regular people can't. Therefore, a supervillain should be as far beyond our understanding of villainy as we are above a villainous ant,* which means a supervillain should be able to do things regular villains can't.

Like making the world a better place.

Yes: this may sound counterintuitive, but the greatest villains long ago stopped being evil for evil's sake, cackling at how malevolent they are as they tie an innocent to some train tracks. Instead, the modern villain is *relatable*. They do what they do with motivations anyone can understand, and they strive to take over the world for reasons that you have likely mulled over while reading the news and savoring a nice cup of tea— reasons that boil down to the simple fact that *you could do it better.* If only those fools would listen, the world really could be a better place. This is

---

*Famously, ants are a collective society—sometimes called a "superorganism"; there's that prefix again—in which members cooperate with one another using a division of labor so advanced that individuals are unable to survive for long without the collective. Therefore, a supervillainous ant would probably be fiercely individualistic, trying to kill the queen and take over the hive. In colonies with more than one queen, this is actually what some queens do to establish dominance, but there's no record of mere worker ants (which are all female, but sterile) trying to take over for themselves. Incidentally, sometimes instead of fighting, ants of neighboring colonies will work together and form a "supercolony" with millions of ant members. In 2009, we discovered that several of the supercolonies of Argentine ants that were introduced by humans to Japan, America, and Europe were in fact part of a single global *megacolony*: when brought back together, the ants from each continent refused to fight one another. This makes ants the most populous society on Earth, besides humans, and thus concludes this really interesting footnote about ants.

what I'll call "enlightened supervillainy," and it happens when someone ambitious, competent, and clever works outside existing power structures to do something that's not getting done on its own. Sounds dangerous? Sounds evil? That exact description could also describe any number of heroes, including all-time champions like Superman, Batman, Wonder Woman, Spider-Man, Captain Marvel, Squirrel Girl, and more.

Everyone likes *those* people!

HOW WE'LL DO IT

It would be easy—too easy—in writing these heists to assure you of my own confidence that they'll work. The end result would be you thinking me a poor deluded egoist, some lost soul high on the fumes of his own imagination. This is fair. So instead, I have chosen a much more difficult but rewarding task. I won't convince you that I could pull off these heists.

I'm going to convince you that *you* can.

This text is your supervillain education, and it begins right now. You started this day as a reader of popular science books, but you're about to learn more about physics and biology and history and technology and computers and space—about the human condition, the universe, and our place in it—than you ever knew before. I'm committed to making you into what you were always meant to be: Someone daring. Someone unprecedented.

Someone supervillainous.

Most crime is petty crime—uninspired, selfish, pedestrian—and I will not be discussing such tedious banalities here. Instead, I want to direct your attention to nine of the biggest, boldest, and most world-changing supervillainous schemes, inspired by comics and pulp fiction—all of which depend *only* on real-life science and technology, without needing even a single comic-book staple like antigravity generators or unstable protomolecules. Each has been masterminded, researched, and designed by me—a trained scientist and professional villainous-scheme creator—to be both scientifically accurate and achievable. And as would befit such wild ambition, none of them are easy. They require strength of character, the ability to see the world not as it is but as it could be, and the resolve and determination to drag that new world, kicking and screaming, into existence. In other words, they require the will, the drive—and yes, the funding—of a supervillain.

That last bit is the catch: you're going to need some money to make all these heists a reality. But fear not! Even if you're somehow unable to secure the funds required for all the schemes in this book (a steal at just $55,485,551,900 USD . . . or less), many of them are individually much more affordable: a decent shot at sending a message through time a thousand years into the future comes in at under $19,000!* But just in case, included in this prospectus of plots is a way to transform an initial investment into an almost-limitless supply of fluid capital (see Chapter 5, "Solving All Your Problems by Drilling to the Center of the Planet to Hold the Earth's Core Hostage").

As you explore the heists and stratagems in the pages that follow, you'll first encounter an appropriately motivational villainous quote, before proceeding into the background required for the heist: think of this part less as "required reading" and more as "casing the joint." This is followed by previous, inferior attempts to achieve the success we're reaching for—made by others not nearly as supervillainous as yourself—before we go over the proposed scheme in detail. I then conclude with an examination of any downsides you should be aware of during the plot—showing you I have done my due diligence—along with any possible repercussions you could face on the off chance you're apprehended by the authorities. Each scheme then concludes with an executive summary, detailing the investment required, the potential return, and the estimated time until heist

---

*If you're curious: as of 2021, precisely 20 people on Earth could afford the full $56 billion investment this book demands, while over a billion people have enough personal wealth to cover $19,000.

maturity. Illustrations by Carly Monardo, an associate of mine, are peppered throughout the text because illustrations rule and authors who keep pictures out of their books are cowards, terrified that their slight words will be upstaged by any proximity to an intuitive, evocative, and honestly more charming visual medium.

You can read these schemes in order, in which case we'll proceed from supervillain mainstays like establishing a secret base and private nation before moving on to more audacious plots to clone dinosaurs, control the weather, and to never, ever die. But if you can't wait to jump ahead, don't let me stop you: chapters are cross-referenced, so you'll be able to quickly find the relevant information you need, and an index at the back is supplied for your villainous convenience.

Now then. Let's proceed, shall we? This world isn't going to take over itself.

Thomas Midgley Jr.

Could one person really accomplish every scheme in this book and produce such an outsized effect on the world? It may seem unlikely, but that's simply because you haven't yet considered the case of one Thomas Midgley Jr., born on May 18, 1889.

In the space of his short life, Midgley was instrumental in inventing both *leaded gasoline* (responsible for poisoning countless humans—including himself, after he poured some of it on his hands and inhaled the fumes at a press conference to demonstrate how safe it was) and *chlorofluorocarbons*, which depleted the ozone layer so efficiently that decades later most countries on Earth were forced to band together and agree to stop making them in order to maintain the habitability of the planet itself. One of Midgley's own inventions did finally manage to kill him in 1944, when he was strangled to death by an elaborate system of ropes and pulleys he'd invented to help get himself out of bed. Nobody has ever credibly suggested that Midgley intended to harm anyone with his inventions, which means all the death and suffering he put in motion during his time on Earth was done by accident. This in turn suggests a question: What could Midgley have accomplished if he had *really* been trying?

PART ONE

SUPERBASIC SUPERVILLAINY

EVERY SUPERVILLAIN NEEDS A SECRET BASE

Three may keep a Secret, if two of them are dead.
—*Benjamin Franklin (1735)*

Every villain needs a place to live, work, and scheme. While civilians may content themselves with a "home," an "office," or a "home office," you're going to live the supervillain dream by plotting in style and comfort from your own palatial *secret base*.

There are some restrictions to keep in mind when scoping out locations for a secret base. The "secret" part of "secret base" means it should be hidden, or at the very least inaccessible: you don't want meddling do-gooders easily stumbling across it. "Base" means it should be sustainable and self-sufficient, able to support you (and ideally a staff of henchpeople) for months if not years at a time. Remember, if you can't bunker down in it for the long term, then you don't have a base: you have a vacation home.

And one simply does not take over the world from a secret vacation home.

BACKGROUND

First, let us dissuade you from what you're already thinking, which is this:

> Obviously the best place for me to build a secret supervillain base
> is inside a volcano, this is easy, I don't even know why I bought this
> book since I know all this stuff already.
>
> —*You (currently)*

Building in an active volcano is a bad idea: it can explode with little to no warning, cooking you alive as the air fills with toxic gases and rocks rain from above onto a floor that is *literally* lava.* Even a dormant volcano means you're living inside a very visible, non-secret hole/tourist attraction.

Above: a plan that has backfired somehow.

---

*That's not to say this has stopped people from doing just that! The small and isolated Japanese island of Aogashima is both shaped and dominated by the volcano that created it, but a population of around 170 people live there anyway, enjoying a windy, rainy climate and natural volcano-powered hot springs (which the locals use to cook and steam food). When the volcano last erupted, in 1785, it killed between 130 and 140 of the over 300 residents who couldn't evacuate in time—but within 50 years the island was repopulated.

Your main concern here is self-sufficiency: if your base is to support you and your henchpeople without having to rely on the outside world, it will need to be a certain minimum size. The precise nature of that minimum size depends on how you answer the question of "wait, how much space do we *really* need to keep a human being alive indefinitely?"

Various authorities have attempted to answer this question throughout history. In the 700s CE in England, land was measured in hides, which reflected the amount of land thought necessary to support a family. Hides ranged in size from around 240,000 to 728,000 square meters (m^2), depending on the productivity of the land,* but around the Norman conquest in 1066 CE, they became standardized at around 485,000m^2: slightly less than half a square kilometer. Whether families at the time included just immediate family members or also extended family is now unclear, but if you assume a small family of just four people, that works out to 121,250m^2 of arable land per person.

Factoring in the modern farming technologies developed over the past millennium, a more recent 1999 calculation determined that a diversified and sustainable European (meat-eating) diet demanded 5,000m^2 of farmland per person, further calculating that if you assumed a largely vegetarian diet; no soil degradation, erosion, or food waste; ample irrigation; *and* godlike farmers who both planted and tended to their crops perfectly, you could probably get that number down to just 700m^2 a person. Lower numbers are better here: they help keep your base reasonably sized *and* have the side effect of making it easier for the rest of the world not to starve to death. That's a good thing, given that the United Nations' Food and Agriculture Organization's measurements of global arable land per person

---

*Of course, at the time hides weren't defined in terms of meters (the metric system hadn't been invented yet—heck, the *imperial* system hadn't even been invented yet). A hide was measured in acres (the word "acres" is derived from Old English, where it means "field"), and one hide ranged from 60 to 180 acres. While today acres have a fixed area (approximately 4,047m^2), at the time they were defined as the amount of land two oxen yoked together could plow in a single day. They were bound by time, not space, which meant the size of an acre—and therefore hide—varied depending both on your soil conditions and on how swole your oxen were.

have been trending downward for decades: in 1970, it was 3,200m² per person; in 2000, it was 2,300m² per person; and in 2050, the global arable land is projected to be down to just 1,500m² per person.

But even these calculations are still just estimates and educated guesses: they're not *facts*. Supervillains ponder and plan and scheme, yes, but they also reach a point where they take bold and decisive action. The supervillain's technique to scientifically discover how much space humans *actually* need to survive is straightforward:

1. Find some humans.

2. Put them in an enclosed area of a certain size, then seal them in so that neither they nor anything else can get in or out.

3. Sit back, relax, and then check in every once in a while to see if your humans died or not.

An illustration of the high-level concept of your experiment.
Don't worry about the one on the right. He's just sleeping.

And even though we're only one chapter in, this book has already saved you lots of money, because I can tell you that this experiment has already been performed before! It was run in 1991 on eight human volunteers during the two-year experiment that was the inaugural run of Biosphere 2, and it cost $250 million USD, equivalent to almost $500 million today. *Money in your pocket, friend.*

Whatever Happened to Biosphere 1?

There were some earlier proof-of-concept prototypes to Biosphere 2, including a sealed test module filled with plants (some to eat, some to produce oxygen), which saw stays by humans that ranged from an initial 72-hour visit to a 21-day experiment in closed-loop, bioregenerative, self-sustaining isolation. However, none of these prototypes were called "Biosphere 1." That's because members of the project considered Biosphere 2 to be a sequel to the natural environment they'd come from, which makes Earth the *true* Biosphere 1. Therefore, the answer to "whatever happened to Biosphere 1?" is "a heck of a lot actually, gosh, *where do I even start?*"

Biosphere 2 is a 12,700m² complex of concrete, steel, and glass built in Arizona whose expenses were privately funded by billionaire Ed Bass. On September 26, four men and four women entered the complex through an airlock, intending to remain for two full years—during which they would depend entirely on the environment inside to keep them alive. In theory, the only thing to enter the Biosphere in that time would be electricity, and the only thing to exit it would be information. The sealed campus was divided into different biomes: a tropical rainforest (modeled on Venezuelan tepuis: tall, flat, and isolated mountaintop ecosystems), a savannah (modeled on South American grasslands), a desert (modeled on coastal fog deserts, with parched ground but moist air), a marsh (inspired by the Florida Everglades), and an "ocean" (salt water, Bahaman sand, and tropical coral reef). Beneath the Biosphere was a basement filled with support machinery, which was also accessible to the "biospherians" inside, since they would be the ones responsible for its maintenance and repair.

Each biome contained its own indigenous plant and animal species: the plants were responsible for producing the oxygen the humans would breathe, while the animals were chosen both for biodiversity and for food. A breeding pair of Ossabaw Island hogs was included due to their ability to "turn almost anything that remotely resembled food into meat and fat," with chickens and goats making the cut for similar reasons: they were sources of meat, eggs, and milk that could eat things that humans won't. Since everything was self-contained, the biosphere had its own water and carbon cycle. The biospherians would, in effect, be drinking the same water over and over again. It was to be the first time in history that humans would be separated for so long from the natural biosphere of their planet.

The experiment was not without its challenges, including:

- The site drew unexpected attention, with tourists gathering and tapping on the enclosures' glass when they wanted a photo of the biospherians. Few places inside the Biosphere actually provided privacy to its inhabitants.

- Some members brought insufficient supplies of clothing from outside, eventually resulting in vital items like boots being barely held together with duct tape.

- Venomous scorpions snuck inside before the Biosphere was sealed, and they had to be hunted by the biospherians to extinction within the Biosphere (there's a word for that: extirpation!)

- Some crops failed, including their only supply of white potatoes, which an infestation of mites forced into extirpation. Sweet potatoes replaced them in the biospherian diet, and they ate so many—half their daily calories came from sweet potatoes alone—that their skin began to turn orange from beta-carotene.

- A species of Australian cockroach stowed away too, and their population exploded before they could be contained and extirpated. Hordes of them swarmed over the floors and tables in the kitchen area at night, turning white countertops brown. The humans countered this by vacuuming them up en masse and feeding them to the chickens, thereby transforming the pest roaches into delicious eggs.

- Despite a carbon dioxide scrubber, CO_2 inside the Biosphere required daily monitoring to make sure it didn't exceed safe thresholds. Grasses in the savannah were cut down and stored in the basement as a sort of manual carbon-sequestering system.

- Oxygen inside dropped to dangerous levels, eventually requiring two oxygen injections into the Biosphere, one a mere month away from the two-year mark. (It was later discovered that the concrete inside the building was sequestering CO_2, which is what was causing oxygen to seemingly disappear from the Biosphere. The concrete was sealed for a subsequent experiment, solving this problem.)

- Twelve days into the experiment, biospherian Jane Poynter lost the end of her middle finger when she accidentally got it caught in a grain thresher. She had to be removed from the biosphere for six and a half hours to visit a hand surgeon.[*]

---

[*]When Jane returned, the project's mission control put a duffel bag inside the airlock with her, including some computer parts and other minor supplies. When some members of the media

And on top of all this, the food was never quite sufficient.

The agricultural land inside Biosphere 2 took up just 2,500m² of space, which works out to a little over 300m² per person: less than half of that best-case-scenario farmland-per-person estimate from 1999. And while the high-fiber, primarily vegetarian diet inside the Biosphere was designed to be nutritionally complete (except for vitamins B_{12} and D, for which supplements were taken), it supplied only about 1,790 calories per person per day: less than recommended levels.* These low-calorie diets caused the women and men inside to lose on average 10% and 18% of their body weight, respectively, during the experiment. At one point, a biospherian calculated that if his rate of weight loss continued, by the time the experiment was over he would weigh *negative 40 kilograms (kg).* Peanuts were eaten whole—the shells at least took up space in the stomach—and some biospherians found consolation in taking turns with a pair of binoculars to spy on a nearby hot dog stand.

During the second year, biospherians resorted to eating emergency grain reserves and excess seeds not intended as food—a short-term solution that meant that if they *were* to stay inside beyond their two years, they might not have enough seeds to plant the following year. Later analysis showed that despite the best efforts of the biospherians—including, at times, gathering to manually pick individual pests from crops in a labor-intensive attempt to save them—the farmland inside was sufficient only for *seven* people, not eight.

---

found out items had been snuck inside, they claimed the experiment had been compromised, but most inside the Biosphere didn't feel the same way, seeing small "cheats" as better than abandoning the experiment entirely. Jane herself ate only one granola bar and drank one glass of water while outside, and she left her lost fingertip inside the biome, writing later, "A small part of me would not be leaving the Biosphere at the end of two years."

*The 2,500m² of agricultural land doesn't include the livestock bay, which was about 135m² big. While there were occasional feasts, meat was never a significant part of the biospherian diet: averaged over the two-year stay, animal flesh contributed just 43 calories per person per day. (In the last three months—with the end in sight—more food was consumed, finally supplying around 2,200 calories a day.)

But perhaps the biggest, and most unexpected, challenge came from the humans themselves. While all eight were friendly upon entering and foresaw no real issues—many were long-standing friends and expected life on the inside to bind them together into an almost utopian society—personality conflicts bloomed and escalated quickly. Within a year, the eight had fractured into two hostile groups of four, with members from one barely speaking or even looking at members of the other group outside of formal meetings. Biospherian Jane Poynter—she of the amputated finger—named the groups: "Us" for the one she belonged to, and "Them" for the other four members. The atmosphere was tense, aggressive, and fraught: at one point two members from Them actually spat in Poynter's face. One walked away without a word, and the other responded to her stunned "What have I done?" with "That's for you to find out." In *The Human Experiment*, her autobiography about her time in Biosphere 2, Poynter wrote, "We were all suffering, hurt beyond words," noting that in the last few months, "I felt the air would ignite, explode in flames with the spontaneous combustion of so much suppressed anger, fanned by a grueling schedule. I had the unnerving feeling that someone would hurt someone at any time." Once they left Biosphere 2, after two years and twenty minutes inside, Poynter wouldn't speak to the four of Them for over a decade. In a 2020 documentary about the experience, *Spaceship Earth*, a former biospherian said, "One of the reasons that our group of eight was belligerent toward each other was we were suffocating and starving." (A second, abbreviated six-month mission at Biosphere 2 with just seven people achieved food self-sufficiency—and added an on-call psychologist.)

Anyway! All that expense, drama, suffering, starvation, and heartbreak allows *us* to conclude that a sustainable base for yourself and six henchpeople, selected for loyalty to you and a predilection not to break off into groups and get all weird and sad, requires at least 12,700m^2 of space, with 2,500m^2 of that dedicated to farming. If you'd like a larger staff, increase the farmland and oxygen levels some and you're golden.

What's a Little Isolation Between Friends?

These biospherian reactions to isolation aren't unprecedented: when stuck together in places with minimal privacy, forced social interaction, and few chances to escape interpersonal conflict, humans can get *a little strange*.

In 1970, at the T-3 station (a small research outpost built on top of a 60-kilometer-square [km^2] iceberg floating in the Arctic Ocean, at the time populated by 19 isolated scientists), a dispute over a stolen jug of homemade raisin wine ended with one crew member fatally shooting another with a shotgun. In 1980, after returning from a stay on the Soviet Salyut 6 space station, cosmonaut Valery Ryumin told his diary, "O. Henry wrote in one of his stories that if you want to encourage the craft of murder, all you have to do is lock up two men for two months in an 18-by-20-foot room. Naturally, this now sounds humorous. Confidentially, a long stay even with a pleasant person is a test in itself." And in 2018, a scientist at Russia's Antarctic station was charged with grabbing a knife and stabbing a colleague he disliked in the chest: early reports said it was over him repeatedly spoiling the endings of books, a hilarious but sadly fictional motivation for murder. (The two later reconciled.)

But there's some success stories too! A NASA-sponsored base in Hawaii (the Hawaii Space Exploration Analog and Simulation) kept a crew of six volunteers inside its $111m^2$ dome for a full year in order to simulate what a mission to Mars might be like. The group did splinter into two ("It wasn't even two cliques as much as two tribes," crewmember Tristan Bassingthwaighte said later. "Soon, all your personal time was spent with people who weren't driving you crazy."), but there wasn't even a *single* murder attempt! Two of the volunteers even started a romantic relationship—which ended just after their year in the dome did, when both were no longer limited to a field of, at most, five other humans for potential suitors. Ouch.

However, true self-sufficiency includes energy self-sufficiency, and these figures don't include power generation. Therefore, we're going to expand your base a bit to accommodate a few small modular nuclear reactors, also known as SMRs. SMRs—defined as small, self-contained nuclear reactors with a limited amount of radioactive material, generating 350 megawatts (MW) or less*—are an emerging technology, and their proliferation has been stymied by regulations (written with larger and more dangerous reactors in mind) as well as by actual nuclear nonproliferation treaties. One design, the NuScale reactor, promises 60MW of power from a single cylinder just 25m tall and 4.6m wide, requiring minimal maintenance and with enough passive safeguards and so few moving parts that it will be "virtually impervious" to meltdown.[†] When an SMR reactor is used up, they can, by design, be easily moved off-site for disposal elsewhere.

But let's be conservative supervillains and limit ourselves to SMRs that have actually been built.

---

*Still, 350MW is a heck of a lot of power! For a sense of scale, a power station operating constantly to generate 1MW of power—1 million watts—is roughly enough to power anywhere from 400 to 900 Western-style homes, depending on their usage.

[†]Fully interrogating how much heavy lifting the word "virtually" is doing in that phrase would take up too much time, so let's not!

In 2019, Russia activated its first self-contained floating nuclear power station: a *nuclear barge*, named the *Akademik Lomonosov*, designed to be moved to whichever areas needed power most.* On board are two KLT-40S SMR nuclear reactors generating 35MW of electrical power each (delivered through electrical wires), and an additional 60MW of heat (delivered via water pipes). They're designed to operate nonstop without refueling for three to five years, and to shut themselves down in case of emergency even without human intervention. The *Lomonosov* is only 144m long and 30m wide, which means adding another 4,320m² to your base's area is enough to accommodate such reactors. Heck, if you want to get even fancier, still smaller reactors have been built for military nuclear submarines: an environment in which a low-maintenance, reliable, and self-sufficient power source is obviously extremely desirable. The smallest nuclear submarine ever launched by the U.S. Navy, the NR-1, was 45m long and 4.8m wide and had just 13 people in its crew.

But we'll stick with the larger, more sustainable, and presumably safer example of the *Lomonosov*, which has a crew of roughly 70 responsible for its operation and maintenance. And yes, that's a lot of new people to feed and water: if this is prohibitive, you could explore solar power, wind power, or simply managing a couple of measly nuclear reactors all by yourself and *just kinda seeing what happens*. If you do go the full-nuclear-staff route, a biosphere 12 times larger than Biosphere 2 would ensure there'd be enough food, air, and water for you, your henchpeople, and your power-generation staff.

Summing up:

---

*The *Akademik Lomonosov* currently floats off the shore of the Arctic port town Pevek, in northeastern Russia, where it began delivering power on December 19, 2019, and was declared fully operational on May 22, 2020.

FOOD SELF-SUFFICIENCY FOR SEVEN PEOPLE	AIR AND WATER SELF-SUFFICIENCY FOR SEVEN PEOPLE	NUCLEAR POWER GENERATION	A FULL NUCLEAR GENERATOR SUPPORT STAFF	APPROXIMATE MINIMUM SIZE (IN SQUARE METERS)	COMPARABLE TO
✓				2,500	A large, very efficient greenhouse.
	✓			10,200	Biosphere 2, only everyone inside has starved to death.
		✓		4,320	A floating nuclear barge.
✓		✓		6,820	A large, very efficient greenhouse that's nuclear powered for some reason.
✓	✓			12,700	Biosphere 2.
	✓	✓		14,520	A nuclear-powered Biosphere 2, only everyone inside has starved to death.
✓	✓	✓		17,020	The penultimate secret base, supporting you and six henchpeople.
✓	✓	✓	✓	144,020	The ultimate secret base, supporting you, six henchpeople, and a nuclear support staff of 70.

A secret grid showing the secret options you have regarding the secret size of your secret base.

That knocking you're hearing is Opportunity—hey, she's there with Science!—and what the two of them are saying is that you actually *can* have a truly self-sufficient nuclear-powered secret base for yourself and 76 of your closest friends, all in less than 15% of a square kilometer: potentially much less, if you incorporate verticality into your lair instead of building it all ranch-style. They're shouting it over the noise of their own knocking; the two of them are *that* excited about this!

Now all you need is a place to build it. Let's see what your options are.

Opportunity knocked.

THE INFERIOR PLANS OF LESSER MINDS

Bases on Land

It's very hard to be secret when you're in plain sight of anyone nearby, and most land on the planet is claimed by one nation or another (see Chapter 2 for a way around that). You will never have true independence or secrecy while being subject to another nation's authority, much less while being monitored by that nation's authorities. This is an attainable but entry-level suggestion for a base, and you can do better.*

A not-so-secret base on land.

If you do decide to pursue a land base, be certain you've built it someplace you can actually access. During the COVID-19 pandemic, American James Murdoch (the son of Fox News media mogul Rupert Murdoch) admitted that his "end of times" cabin in Canada, constructed on a 445-acre lot with its own water supply and solar panels, was inaccessible to him. "The borders got shut so I haven't been," he said. "I don't know why we didn't think that through."

Underground Bases

Going underground affords you the chance that your base will remain hidden over the long term. But moving all that dirt without being noticed is

---

*Until late 2020, RiotSheds.com offered high-carbon-steel-armored, pistol-resistant, riot-proof sheds for sale, designed to look as boring and uninteresting as possible during periods of civil unrest: an option for those who want just a taste of the secret-base lifestyle. They cost $29,999 USD, and getting one with a bench or gun rack inside added an extra $208 or $180 to your bill respectively, but delivery *was* free within the lower 48 American states. Unfortunately, they seem to have gone out of business while I was revising this book. It's too bad! Look at all this free footnote advertising they missed out on!

still going to be difficult (see sidebar on page 19), and farming without sunlight is a challenge at the best of times. Many survivalists have nevertheless pursued this option, only they call them "bunkers" instead of "bases," and envision them more as someplace to retreat to in the immediate aftermath of a civilization-ending disaster, where they'll rely on a limited supply of dry and canned goods for food instead of striving for true self-sufficiency.

If for some reason you don't feel like building your own bunker from scratch, there are several corporations offering ready-to-go apartments as part of exclusive—and very expensive—underground survival complexes. One such example of a "billionaire bunker" is Survival Condo, a 5,000m² complex of 14 luxury residences built into an old missile silo in Kansas, itself originally constructed to withstand a nuclear blast. There's a gym, a bar, a library, a classroom, a spa, a cinema, a 180,000 liter swimming pool, a lounge, a hydroponic garden, a dog park, a climbing wall, air-hockey tables, three armories, a shooting range, and five years' worth of stockpiled food. Gaining access to all these amenities through buying a single-floor furnished suite will only set you back a cool $3 million USD. (This does, however, suggest a very easy, and very villainous, way to survive for at least a few years underground during a global catastrophe: simply sneak your way inside one of these communities before anyone else, then lock the doors behind you. Instant villainy, instant millions of dollars in savings, and instant access to a whole nuclear silo's worth of dried goods.)

But even if you were comfortably ensconced in such a decadent bunker, and you'd somehow converted large parts of it for farming, you'd still be living with the weakness that if the power ever goes out, you're going to be thrust into total darkness. This is obviously deadly for crops, but, perhaps surprisingly, it's also suboptimal for humans too. In 1962, Michel Siffre, a French caver, intentionally spent two months in a dark glacial cave to see how it affected him. With only his flashlight as a light source, he found that he had no way to mark time except with his own body. He soon lost track both of time (his count of what day it was was 25 days short) and of his own past. "When you are surrounded by night . . . your memory does not capture the time. You forget. After one or two days, you don't remember what you have done a day or two before . . . It's like one long day," he said in an interview years later. When he repeated the experiment a decade later, this time spending over half a year in a Texas cave, he reported thoughts of suicide, 80 days in, when his record player broke.

A secret underground base.

You can, and should, dream bigger.

Some New Digs

While building underground bases is hard, it's not impossible! In the late '50s, the U.S. government began "Project Greek Island," an ambitious 10,400m² secret base hidden *underneath* a West Virginia hotel, which used construction from that hotel's remodeling as cover. It was designed to house members of both the House of Representatives and the Senate, allowing them to live and legislate from an undisclosed but safe location in the event of nuclear war for up to six months at a time. But in 1992, it became useless for that purpose when *The Washington Post* discovered and reported on the project, after which the facility was decommissioned. So much for that!

The private sector has had some successes too: in 1975, a former insurance agent named Oberto Airaudi moved to the Piedmont valley in northern Italy with his then-24 followers and renamed himself Falco. Together they began building a commune, including a secret base/worship site 30m underground named the Temples of Humankind. Buckets of gravel and dirt were carried out by hand and disposed of quietly, and as construction progressed, a surface-level decoy shrine was built to defang rumors of a secret temple complex being built outside of town. The Temples of Humankind evolved into a five-story, 8,500m³ complex—complete with 12m-high gold-leaf ceilings, bronze statues, and giant frescos.

The commune itself was (and is) focused on self-sufficiency, having created their own constitution, currency, schools, tax code, and newspapers; however, all food is still grown above ground. Members wishing to become full "Class A" citizens sign over all their worldly possessions to the commune, a practice that went well until 1992, when a former member left and wanted his stuff back. He sued for it, and in so doing made the existence of the underground temples public. Shortly thereafter, Italian police arrived to shut down the illegal dig site but couldn't find the entrance to the underground complex. Only after they threatened to dynamite the whole hill to find it did Falco bring them inside, and the Italian authorities found the complex so impressive that they *retroactively* issued a building permit for the site. Today, three-hour tours of the underground base are offered by members of the 600-strong commune for €70.00.

And if I could be so bold as to direct your dreaming somewhat: Wouldn't it be amazing if your base were *mobile*? This way if it were ever in danger of being discovered or threatened with attack, you could simply move it someplace new. And since we're dreaming, let's imagine this place your base is moving to recognizes no nation's authority, is almost wholly uninhabited, and is already mythologized by countless generations as someplace vast, unknowable, dangerous, capricious, *and* deadly. In fact, let's give it such an effective reputation for death that the rest of the world has long ago accepted that sometimes people are going to enter this region and disappear and their bodies will never be found and *this is just an unremarkable thing that happens there sometimes.* Heck, while we're wishing on a star, let's also demand that for millennia those who have been to this area and returned alive carry with them tales of terrifying and impossibly monstrous beasts that attack without warning and tear humans apart.

Hey, that brings us nicely to the next section!

Sea Bases

The ocean is a tempting venue for a secret base. Over 70% of the Earth's surface is water (which means you've got lots of locations to choose from), and most of it is underpopulated international waters (which means you're free from any nation's authority, laws, or supervision). Check it out: we're only two sentences in, and already the ocean is looking great!

A logical starting point for your secret sea base is cruise ships, which are often conceived of—and marketed as—floating cities. But they're actually closer to floating hotels, inhabited by a largely transient population of tourists supported by a full-time staff focused on their comfort. Perhaps the closest thing to what we want is *The World*: a 196m-long cruise ship, with 12 decks of about 5,000m² each, marketed toward the ultra-rich as the largest private residential ship on the planet.* Instead of rental accommodations, each of its 165 residences is privately owned, condominium style: suites cost between $1 million and $8 million USD to buy (not including upkeep), and every owner must prove they have a net worth of at least $10 million USD. The focus on board is on "anticipatory service and bespoke experiences": lectures by Nobel laureates, a wine cellar with 1,100 different wines from 19 countries, cocktail lists constantly adjusting to reflect the culture of the ship's current location, so many restaurants sourcing food from exotic ports of call that they open on a rotating schedule to ensure there's sufficient clientele to avoid making them feel like ghost towns, a full-sized movie theater named Colosseo, a gyroscopic and self-leveling pool table, library, spa, *and more*. But even this boat, for all its size and wealth, is not self-sufficient: it needs regular stops to resupply the millions of dollars' worth of fuel, food, and entertainers it demands as it endlessly circumnavigates the world for the pleasure of its (pet-free, median age mid-sixties) residents. And when the COVID-19 global pandemic hit the rest of the world in March 2020, *The World* was taken out of service, its secret clientele of rich elderly people unloaded and brought to rest in a berth on the southern coast of England.

It's worth stressing that last point: even though nobody on board had contracted the novel coronavirus, its residents were *still* forced to join the rest of the world in their reality of pandemic and quarantine. This is of

---

*This record may be challenged in the near future when its competitor, the 296m-long *Utopia*, launches. Residences on board range from around $4 million to $36 million, and it too is marketing itself as an exclusive and private full- or part-time residence at sea for the richest people on Earth.

course suboptimal for a secret base, but it's the nature of any community that's not self-sufficient: relying on outsiders for resupply means you're regularly subject to the same issues, laws, *and* diseases as they are. The best this ship can offer is a secret base without its own air or water cycles, with a very limited supply of fuel and power, and with only enough room for farming on its sun-soaked top deck to support 14 people at most.

We're gonna need a bigger boat.

What you want is a ship so colossal that it can actually *be* self-sufficient, and reliable enough that if a pandemic hit, you and your staff could simply float out into the ocean—far from the madding and/or disease-carrying crowd—and wait until it passed. There have been various proposals for such ships, including the name's-a-little-on-the-nose *Freedom Ship*: a titanic 1.4km-long vessel (actually, a joined set of eight flat-bottom barges working together as a single vessel, providing around 150,000m^2 of space on each deck) that would function as a city. There would be condo-style residences, hotels, a hospital, schools, a transit system to get from one end of the ship to the other, and the world's largest duty-free (and tax-free) shopping mall. The boat itself would be big enough to accommodate an airstrip on the roof of the vessel, and 100,000 people could be on board at once: 40,000 full-time residents, 10,000 hotel guests, 20,000 full-time crew, and 30,000 day visitors. But since the project's conception in 1999, *Freedom Ship* hasn't come close to being built, and its website today—while still bullish on construction starting soon—makes money by selling branded print-on-demand mugs, stickers, and thong underwear on CafePress.com. And less-ambitious proposals—like French architect Jean-Philippe Zoppini's *AZ Island*, a 100,000m² floating town—haven't done much better: that one got as far as being announced with a shipyard attached to build it as soon as orders came in, but it fell through due to a lack of funding.

Living the Sweet Life

AZ Island was to be an oval-shaped floating island/hotel/city complex inspired by Jules Verne's 1895 novel *Propeller Island*, wherein—and this may sound familiar—a massive city-sized ship is built and then inhabited entirely by millionaires. For some reason, publicity for the project didn't focus on the conclusion of Verne's book, wherein the vessel falls apart and sinks with a massive loss of life.

Verne himself may have been inspired by the 1833 prospectus by John Adolphus Etzler, titled *The Paradise Within the Reach of All Men, Without Labor, By Powers of Nature and Machinery: An Address to All Intelligent Men, In Two Parts*. In it, Etzler presented what claimed to be "no idle fancy": floating islands, homes for thousands of families made from wooden logs, powerful engines fueled by the motion of the waves themselves, all held together by living trees "reared so as to interweave each other and strengthen the whole." They would be covered with fertile gardens and palaces and comforts and luxuries, and would move across the oceans at 60km/h, "exempt from all dangers and incommodities." Etzler claimed an investment of $300,000 USD then ($9.6 million today) was all he needed to accomplish this within a decade, transforming the world into a luxurious paradise "surpassing all conceptions." Despite proposing a public share drive (shares costing "not greater than the price of a lottery ticket") and sending a copy of his book to President Andrew Jackson to lobby for funding, he never got the money.

Etzler himself disappears from the historical record after starting a utopian community in Venezuela, where he promoted a life of leisure free from disease, where settlers would feast on pure sugar: "the most important article of food" that, he believed, was spontaneously deposited onto the ground by the atmosphere itself within sufficiently warm climates. He attracted 41 settlers, and 15 died within five months. A report written afterward declared, "Never, probably, did any body of European immigrants . . . find themselves in such a state of destitution and unmitigated misery." So it was probably for the best Etzler wasn't put in charge of floating city-states after all.

Even without *Freedom Ship*'s $10-billion-plus price tag, there are still some challenges in how it could be built, maintained, and kept seaworthy over the long term. Ocean water (and salty sea air) is corrosive, and there's no dry dock in existence that could handle a ship (or series of ships) of this scale. UN international standards have ships in dry dock at least twice every five years for hull inspections, which—if you choose to follow them—means more land-based infrastructure will have to be both built and maintained to keep your base going. As for self-sufficiency, *Freedom Ship*'s top deck would have the equivalent of 11 biospheres of area, which could produce enough air and food for 77 people in a completely sealed environment, or 420 of them if you grow food but don't recycle your air and water.

But perhaps we can relax our requirements even further. At sea you could rely on the ocean's bounty for food, and solar and wind power for electricity, couldn't you? What's to stop you from building a much smaller and simpler floating base on the ocean and living there indefinitely, free at last from both land and its associated lubbers?

A seacret base.

Seasteading is a utopian- (and libertarian-) tinged movement with the goal of doing just that: living on the sea, free from any and all government, laws, and coercion, in a state of true and radical freedom. The closest that's come to reality so far is with a small single-family seastead installed some 20km off the coast of Thailand in 2019, the first of a then-planned fleet of identical residences. The 6m-wide, two-story octagon-shaped structure was launched without the knowledge or permission of the Thai government. "We have been keeping under the radar so far but we follow all the laws of Thailand so it's as if we're just living on a boat in the water as far as they're concerned," seasteader Chad Elwartowski told the magazine *Reason*, shortly after he and his partner, Nadia Summergirl, began living on board. But within weeks of moving in, the Thai government noticed them and characterized their floating house as:

- a threat to ocean navigation,

- an illegal occupation of sovereign territory,

- a treasonous attempt to form an independent state in their national waters, and

- an attack on Thai national security.

The seasteaders, now wanted criminals under Thai law, fled shortly before the navy commandeered and destroyed their seastead, bringing with them only a few belongings—and a treason charge that, if they are ever captured and convicted by the Thai authorities, carries the death penalty.*

---

*While on the run from Thai authorities, Elwartowski and Summergirl claimed to be just "enthusiastic supporters of the project who were lucky enough to be the first ones to stay on it," and that the company actually responsible for the seastead's design and construction was a mysterious organization called Ocean Builders. After they'd gone into hiding, Patri Friedman, a friend of the couple and the chairman and cofounder of the Seasteading Institute, answered a CBC interviewer's question of "[Elwartowski and Summergirl] didn't design or pay for [their seastead]. So who actually was in charge of this boat?" with "It's a mystery." After escaping Thailand and moving to Singapore, the two now live in exile in Panama, where, according to *The Guardian*, they've relaunched their Ocean Builders company with the intention of building new and improved seasteads off Panamanian shores. Ocean Builders' website refers to what happened as a "successful prototype in Thailand."

So, again: suboptimal.

Other attempts at building floating cities—this time with the full involvement and permission of nearby countries—have run into similar trouble. The Seasteading Institute (an organization funded in part by PayPal cofounder Peter Thiel, who wrote in a 2009 blog post praising seasteading that "I no longer believe that freedom and democracy are compatible")* has been exploring living at sea since 2008. After determining that it's prohibitively difficult and expensive to build and live in international waters far from shore, the Institute negotiated with the government of French Polynesia in 2016 to build a prototype for their Floating City project in the Atimaono lagoon off Tahiti. It was to be a modular and semiautonomous experiment in self-sufficiency (and self-governance) at sea, powered by wind and solar energy and exempted from taxation by the French Polynesian authorities. Plans called for the city to be built out of a series of joined platforms, each one costing between $15 million and $30 million to construct—with the caveat, of course, that if any individual platform owner didn't like how things were going, they were free to detach and float somewhere else.† However, after protests by Tahitian residents—upset that their lagoon was being colonized and polluted by outsiders who wouldn't even pay taxes for the privilege—the government declared in 2018 that the agreement they'd made was not actually a legal document or contract, and besides, it had expired back in 2017.

Even significantly less ambitious projects than full-scale floating cities have run into fatal obstacles: a 2011 scheme by the Seasteading Institute to

---

*In that same essay, Thiel also wrote, "I stand against confiscatory taxes, totalitarian collectives, and the ideology of the inevitability of the death of every individual." For more on *that*, see Chapter 8: "How to Become Immortal and Literally Live Forever."

†More ambitiously, there were (and are) seasteading visons of a "marketplace of governments," where governance is a product like any other—leaving citizens free to shop around for the one they like best, moving (or floating) as they wish between, say, socialist seastead communities, libertarian ones, and ones ruled by religion. This also opens the door to seasteads forming experimental "startup constitutions," "startup countries," and "*Bioshock*-style startup authoritarian regimes."

get around immigration laws and give American companies access to non-American tech workers by having said workers live and work visa-free on a giant ship anchored outside American territorial waters collapsed when funding failed to materialize, and this was after a decommissioned cruise ship, the 67m-long *Opus Casino*, was donated to the Institute.

How Much Water Off Their Shores Does a Country Control?

As defined by the United Nations, there are several different classes of territory in the waters off a country's coast. Territorial sea waters are the closest, running from shore to 12 nautical miles (about 22km) out. This is considered the sovereign territory of the nation. After that comes the contiguous zone, running an additional 12 nautical miles out, where states have less control but can still enforce laws relating to immigration, smuggling, and so on. And finally, running from shore to 200 nautical miles (around 370km) out is the exclusive economic zone: these waters are reserved for the benefit of the nation at their shore, and while you can't fish or mine them without permission, there are no laws (yet) against loitering indefinitely there in a cheeky floating house.

However, there *have* been successes, at least on a personal scale. Since 1992, Catherine King and Wayne Adams have been living in Freedom Cove: originally a single floating house, now a self-made artificial floating island of 12 platforms supporting their residence, open-air gardens, four greenhouses, an art gallery, a dance studio, a candle factory, two boat-houses, *and* a fire pit, all floating off the coast of British Columbia, Canada, in an area accessible only by boat. The entire 1,600m² complex is hand-built, initially out of scavenged wood, and now out of more robust materials that have either been donated by friends of the couple or traded for with sales of their art and sculpture. Both King and Adams eat food from their garden, with Adams supplementing his diet with whatever fish, crabs, and prawn he can catch—including from a Plexiglas trap door that opens in their living room floor, allowing fishing from the comfort of the couch. (King, a vegetarian, doesn't partake.) With fresh water gravity-fed from a nearby small waterfall, and power coming from both solar energy and a backup generator, the two have been living there—successfully and sustainably—for almost 30 years. Adams makes the 30-minute trip into the nearest town infrequently, to sell art, get mail, or "when I want pop, chips, and candy." But there's constant work to be done, both for regular upkeep and repairing damage caused by storms. In 2019, King told the *Exploring Alternatives* YouTube channel, "It's hard work, and you have to do the work. If you don't do the work, then you can't live this way," with Adams adding, "We don't recommend everyone do this. We recommend everyone do their own dream." The couple will soon need to rebuild their residence, as water has been rotting it from the bottom up.

So we seem to be at an impasse. At one extreme, artists have found small-scale living at sea viable, but only in a protected cove, with a nearby free source of drinking water, and with two people at work so consistently that there's sometimes little time for art, let alone villainous scheming.* But at the other extreme, even the resources of some of the world's richest people have thus far failed to make living self-sufficiently on the ocean's surface a viable proposition. Neither of these offers a lot of hope for our vision of a floating secret base.

---

*Adams and King have also never tried to hide their base, never declared their independence from Canada, and have kept paying their taxes.

But perhaps we're thinking too inside-the-box. Perhaps we are not pushing nearly enough envelopes. Perhaps, instead of living on the sea, we should simply live . . . *under* it?

Underwater Bases

Nope, turns out this is way more difficult! The added secrecy is great—80% of the ocean's floor has never been explored, mapped in any detail, or even photographed, and once you're deep enough underwater, even satellites can't spy on you—but the challenges are prohibitive.

Underwater living combines all the downsides of living at sea with the added danger that stepping outside your base without adequate protection results in you drowning, being crushed to death, *or both at the same time.* The environment is constantly trying to kill you, sunlight is limited or nonexistent, air must be recycled or brought in from outside, and forget about underwater farming: there's no crop that can be grown on the sea-floor that's as nutritious as the crops grown on land, benefiting as they do from thousands and thousands of years of human-led selective breeding. The largest underwater animals we've ever domesticated are carp (we bred them into goldfish and koi, starting in China around 1000 CE), so *also* put out of your head any visions of killer whales pulling plows as you ride great white sharks into glorious battle, no matter how well that would fit in with your personal supervillain aesthetic.

In 1963, explorer and conservationist Jacques Cousteau, his wife Simone, their parrot Claude, and four other oceanographers did survive for 30 days in Précontinent II, a collection of small underwater bases referred to as a "village." But this experiment still relied on surface-supplied air, food, and electrical power; was relatively cramped; and the main living structure was only 10m beneath the surface. And in 1964, the U.S. Navy's SEALAB

---

*And this isn't even a full accounting of the threats! If you swim to the surface too quickly, for example, you could also die from the bends, which happens when depressurized gas *boils out of your blood.*

habitat was lowered a deeper 59m into the water off the coast of Bermuda, where divers stayed for 11 days. (The sequel base, SEALAB II, saw a 30-day stay, but the threequel SEALAB was scrapped after going millions of dollars over budget.) Today, there are underwater research stations—NASA's NEEMO research station, 19m underwater off the coast of Florida, is used to simulate spacewalks—and there's even publicly accessible underwater restaurants and hotels, but none of them are self-sufficient, nor are any of them large enough to *become* self-sufficient. These are places you visit, not places you live in.

Unfortunately, life is not necessarily better down where it's wetter.

But despite us already eliminating both land and water (and the mysterious depths of each) as desirable candidates, there is still one place on Earth open to you.

You, my friend, shall build your base *in the sky*.

YOUR PLAN

Just like international waters, there's international airspace where no one nation's laws apply. And even when you do find yourself flying above land already claimed by someone, if you go high enough, you can usually find airspace that the countries below can't or won't police, allowing you to escape the bureaucracy and state control that plague lesser pilots. An example: in America, when you're flying anywhere between 5.5km to 18.3km above sea level*—this is where most planes fly—your aircraft must obey instructions given by air traffic control at all times *and* have been given prior explicit clearance to enter said airspace. But escape to the airspace *above* 18.3km and you're classified differently: there, no radio communication with or clearance from air traffic control is required. It's airspace that historically hasn't been regularly used for much of anything, which means it's *wide open* to our purposes.

So, great: we can soar through the skies above America as much as we desire, so long as we stay 18.3km or higher above sea level. Now all we need is to pull off the ancient human dream of perpetual, indefinite, limitless flight. And the good news is, we're not as far away as it might seem!

Some Exciting News About (Air)space

If you own land in America, you may be interested in the case of *United States v. Causby*. Causby, a farmer, sued the U.S. government over bombers and other airplanes flying above his farmland. When these planes landed at a nearby airstrip, the sudden noise startled his chickens so much they *killed themselves*, with up to 10 chickens every night snapping their necks against the walls of his barn in their terror-fueled flailing. The case went all the way to the Supreme Court!

*The odd number is because of America's use of the imperial measurement system: 18.3km corresponds to 60,000 feet. To put a height of 18.3km in context, humans can't live at this height without a heated and oxygen-pressurized capsule, and if you fell, you'd have several minutes to mull over the choices that led you to that moment before you hit the ground while traveling at over 200km/h.

In its 1946 ruling, the court rejected the farmer's argument that he owned everything above his land to the edge of the universe, declaring that the thirteenth-century legal principle he'd invoked of "Cuius est solum, eius est usque ad coelum et ad inferos" (Latin for "whoever owns the soil, it is theirs all the way to Heaven and all the way to Hell") had "no place in the modern world." But they also rejected the government's argument that they owned all the airspace right down to the ground, noting that some airspace was needed to erect property, grow trees, or even run a fence. Their ruling resulted in landowners having exclusive rights to the airspace above their land from the ground up to 365 feet above the tallest structure on it, which is a little over 111m.

Humans first flew with hot air balloons, which operate on a simple principle: keep the air inside the balloon lighter than both the air outside it *and* the weight of the basket and humans attached to it, and you'll fly . . . for as long as that remains the case. Unfortunately, since air cools and helium leaks, balloon flights are limited either by their supply of fuel (used to maintain or increase buoyancy) or ballast (used to throw overboard onto surface-dwelling fools, thereby decreasing weight and prolonging flight). The longest balloon ride in history was accomplished with a combination helium-and-propane-powered hot air balloon named the *Breitling Orbiter 3*: in 1999, it became the first balloon to fly around the world without landing, setting a record flight time for any unrefueled aircraft by staying airborne for 19 days, 21 hours, and 47 minutes.* But even the pilots

---

*The record still stands as of the printing of this book. The balloon relied mainly on the jet stream for propulsion, which required flying some 10km aboveground to stay within it. (When

of the balloon (Bertrand Piccard and Brian Jones) credited luck with their success: the luck of the wind blowing them where they needed to go, the luck of their fuel not running out before they landed, the luck of them not perishing when they came within an hour of death from carbon dioxide poisoning after their oxygen source failed in their pressurized capsule, the luck of not being shot down when they crossed into Yemeni airspace without permission *and* when they skirted the edge of Chinese airspace from which they had been explicitly forbidden, and, finally, the luck of them not crashing into the ocean after a buildup of ice made their balloon too heavy to fly, which is exactly what happened to a competing balloon team attempting their own circumnavigation at the very same time.

That's not even a month of continuous flight in a balloon, and that's with everything going right.

Balloons Within Balloons

Breitling Orbiter 3 was a Rozier balloon, which is really two balloons enclosed in a single envelope. One filled with lighter-than-air gas (like helium) provides an always-on source of most of the lift, while another filled with air is heated or allowed to cool to adjust the vessel's overall buoyancy. This drastically reduces the fuel required compared to just using hot air for lift, which of course unlocks longer flights.

These double balloons are named after aviation pioneer Jean-François Pilâtre de Rozier, who invented them as part of

empty propane tanks were jettisoned as ballast, the pilots first reduced altitude until they could see the ground, in order to be certain they weren't putting any sky murders into motion when they dropped them).

his 1785 attempt to fly from France to England across the English Channel. His design used explosive hydrogen gas instead of helium though, and half an hour into the flight—in a twist of fate that now seems inevitable to we post-*Hindenburg* readers but which *probably* came as a surprise to him—the balloon caught fire, the hydrogen ignited, the vessel crashed some 1.5km to the ground, and Rozier and his companion became the first two people in history to die in the crash of a (viable) aircraft.

It must have been gruesome: a contemporary account published in *The Derby Post* describes them as ". . . dashed to pieces. I was with the bodies in half an hour and never saw anything so shocking," adding, "I hope to never hear of another [air crossing] being attempted in this or any other country," and concluding by warning of the hubris of balloonists and their "too-soaring ideas" that fail to respect the dangers of air, "an element as fickle as it is unknown." But don't worry! *You'll* probably be fine!

Planes are even worse: the longest fueled plane ride lasted just nine days and three minutes, when Jeana Yeager and Dick Rutan piloted their specially designed plane *Voyager* around the world in 1986. If you allow the cheat of in-air refueling, the longest flight was the 64-day journey of a small Cessna aircraft in 1958, flown endlessly in circles above the desert to gain publicity for the Hacienda, a since-imploded Las Vegas hotel/casino. But again, in-air refueling—accomplished twice a day by the plane flying just 6m above ground while a chase truck matched speed and hoisted a hose skyward—does not provide either secrecy or freedom from land. And staying in a small plane for two months isn't comfortable either: years later, one of the two pilots of the record-setting flight, John Cook, said, "Next time I feel in the mood to fly endurance, I'm going to lock myself in a garbage can with the vacuum cleaner running. . . . That is, until my psychiatrist opens for business in the morning."

But what about solar power—couldn't that let us ditch having to refuel entirely? *Possibly.* The closest thing we have to a solar-powered plane capable of perpetual flight is the *Solar Impulse 2*, a single-pilot solar-powered plane. During the day it gains altitude and charges its batteries, and overnight it relies on those batteries to fly while slowly losing altitude. In 2015, the plane *did* fly around the world, but it took 17 separate hops, with the longest flight lasting 117 hours and 52 minutes going from Japan to Hawaii.* Given the right circumstances, the right weather, and the right luck, the plane could, in theory, fly indefinitely, but it has so far not demonstrated anything longer than five days straight of flight time. And it's a rough five days: *Solar Impulse 2*'s crude autopilot could do little more than ensure the plane was headed in the right direction, which meant that the pilot, inside its tiny cockpit, could sleep for 20 minutes at a time at most. Plus, there's still the issue of what you're going to eat: none of these vehicles are large enough to sustain even one person long-term.

Smaller unoccupied autonomous vehicles (UAVs) have pulled off longer flights—you can slip the surly bonds of Earth, dance the skies on laughter-silvered wings, sunward climb and join the tumbling mirth of sun-split clouds and touch the face of God a *heck* of a lot easier when you don't need to lug any pesky humans and their heavy food and oxygen around—but a secret base isn't much use if you can't visit it, and none of these UAVs can carry you. So while it may soon be possible to store some lightweight items

---

*This also set the record for the longest *solo* flight for any type of airplane—which still stands, in part because the authority keeping track of these records, the Fédération Aéronautique Internationale, stopped recognizing new endurance records for crewed flight due to the obvious safety concerns involved. André Borschberg and Bertrand Piccard swapped the role of pilot with each hop; Piccard was also on the *Breitling Orbiter 3.*

in endlessly flying *secret storage lockers* 18.3km above ground, we're far from being able to hide out there indefinitely.

So if planes are out, UAVs are out, and balloons are out, and none of them are large enough to support you and the air and water and farmland and henchfolk you require, is all hope lost?

Not necessarily. You still have one last physical property of the universe on your side, a secret weapon discovered by Galileo Galilei in 1638: the powerful, nigh-unstoppable *square-cube law.*

The square-cube law describes how when an object grows in size, its surface area increases with the *square* of the growth factor, but its volume is *cubed*. In other words, when things get bigger, the space inside them grows much faster than their surface area. And this means that a balloon, sufficiently scaled up, reaches a point where the volume of air inside it is so huge that the material used to hold it in becomes almost a rounding error. In his 1981 book *Critical Path*, architect Buckminster Fuller revisited his 1958 proposal for floating cities with a design he called Cloud Nine, based on this consequence of the square-cube law. The basic concept was simple: you need only take one of his patented geodesic domes (they had already been invented in 1925 in Germany, but he was awarded the patent in America in 1954) and pair them with their exact duplicate on their other side, forming a geodesic *sphere*.

And then you make them really, really, *really* big.

A Fuller Life

Buckminster Fuller lived his life as, in his own words, "an experiment, to find what a single individual could contribute to changing the world and benefiting all humanity"—strong supervillain vibes from Bucky there. And he was no stranger to engineering megaprojects! He was an early supporter of Biosphere 2, and one of his former assistants, Peter Jon Pearce, designed its dome-inspired structure.

In the 1960s, Fuller dove into exploring the possibilities of sea bases by drafting his own design (and scale model) for a floating neighborhood named Triton. It was a single high-density 5,000-resident platform with residences, stores, offices, and schools all contained within one colossal building. These floating "neighborhoods"—195,000m² of habitable space built on top of a 16,000m² footprint—could be joined together to form cities of up to 125,000 people. While Triton was designed to be anchored near land indefinitely as a complement to existing metropolises, Fuller noted they could be moved when desired, and that "floating cities pay no rent to landlords." Charmingly, the exterior walls on the Triton project were sloped, with the idea that anyone who fell from a window or balcony might be spared a fatal impact with the ship's deck by instead being rolled out to sea.

Fuller calculated that a geodesic sphere 30m in diameter would weigh about 3 tons and enclose 7 tons of air. And then he started doubling the size of that geodesic sphere, reevaluating the materials required each time, to produce the following estimates:

DIAMETER OF GEODESIC DOME	WEIGHT OF DOME	WEIGHT OF AIR ENCLOSED IN DOME	RATIO OF DOME WEIGHT TO AIR WEIGHT
30m	3 tons	7 tons	0.42
60m	7 tons	56 tons	0.12
120m	15 tons	500 tons	0.03

Just an innocent collection of numbers that law enforcement doesn't need to concern themselves with at all!

As you can see, as the geodesic sphere gets larger, the weight of the structure itself—compared to the weight of the air it encloses—becomes negligible. A geodesic sphere a half mile in diameter (about 800m) would, he calculated, have so much air inside—and weigh so little comparatively—that if the air inside were heated by sunlight to just half a degree centigrade or so more than the surrounding air, it would fly. (This technique of capturing sunlight for heat is what lets greenhouses stay cozy even in winter.) Double the sphere in size to the full mile—1.6km in diameter—and the weight of humans and their structures inside would become as negligible as the weight of the structure of the sphere itself, allowing many thousands of people to live and work inside these floating cities. Fuller proposed that these spheres could either float freely around the world or, alternatively, be anchored to mountaintops. Small aircraft would allow people to come and go as they pleased between floating cities, or move from city to ground.

A supervillain base in the sky. Thank you, Buckminster Fuller, recipient of the Presidential Medal of Freedom, for this idea.

The whole idea both sounds *and is* unbelievable, but the math actually checks out: a 1.6-kilometer-diameter dome would enclose 2.14km³ of air. The density of air changes with pressure and humidity, but dry air at 20°C—room temperature—has a density of around 1.204kg per cubic meter. If we kept the *inside* of our sphere at that temperature, and the air outside was just half a degree cooler, at 19.5°C, the density of that colder air—1.206kg per cubic meter—would be enough to provide over 4 million kilograms of lifting force to the sphere. If the temperature difference were even greater—say, if we floated around the South Pole, where we're all but guaranteed cooler temperatures throughout the day*—then the buoyancy

---

*The highest temperature ever recorded in Antarctica was 20.75°C in February 2020—the first time any temperature above 20°C was recorded, an event scientists on the continent described as "incredible and abnormal"—but the mean annual temperature of the Antarctic interior is below −50°C. Antarctica is also home to the coldest temperatures ever measured on Earth: −93.2°C. All this to say: *you'll generally be fine,* though maybe keep your dome at a slightly warmer 21°C instead and don't dawdle on building your Antarctica secret base for too long, because climate change isn't going to wait for you (though, see Chapter 4 for your options there). Also keep in mind that just heating air inside your sphere won't be enough: you need to

gained would get even better. In air that's just 10°C, our 20°C geodesic sphere would provide over *90 million* kilograms of lifting force. That's (literal) tons of weight, and tons of space: a platform across the middle of a 1.6km-diameter sphere would provide over 2 million square meters of area: plenty of room for any nuclear-powered secret base we're envisioning, and the heat those generators provide. (Such a platform is actually enough room for *thirteen* of those bases, though of course the more things you add inside the sphere, the more you both increase your weight and displace air. You could always place your living platform farther down in the sphere if you don't need all that extra space: a platform 30m from the bottom of your 1,600m-diameter sphere is just large enough for a single base.)

There are limits on how high you can go, of course: air density decreases as you go up, which means your city will reach a point where it'll stop rising and start hovering, and you won't be able increase the temperatures inside any more without making it too hot for the humans to be comfortable.* But your sphere could be refined past even Fuller's design by including, like the *Breitling Orbiter 3*, some internal helium balloons to provide a constant source of lift beyond even hot air.

Now all you need to do is build it. And hey, that brings us to the next section!

---

be able to vent some gases outside, so that this heated air can become less dense and not just more pressurized.

*If you want to reach those mountaintop heights Fuller envisioned, the air in your sphere will need to be less dense than the air a few thousand meters up—where it's already so thin that humans have trouble breathing. You can solve this by keeping a small slice of your sphere pressurized and comfortable for humans, or by just keeping buildings pressurized and having everyone wear oxygen masks while working outside them. And remember too that since hot air rises, the temperature at the bottom of the geodesic sphere—presumably, where you'd do most of your living—will be cooler than at the top. When we calculate the air density at 20°C inside the sphere, we're taking the *average* temperature throughout the sphere, but it'll always be hotter at the top and cooler at the bottom. This works out well for your floating base: you can increase temperatures to get even more lift from the top of your sphere while your habitat still remains comfortable at the bottom.

THE DOWNSIDES

A 1.6km-diameter geodesic sphere would be almost twice as tall as the Burj Khalifa in Dubai—currently the tallest structure ever built by humans—so obviously there are *some* construction logistics that still need figuring out.* But that's not even the hardest part! In truth, the actual largest challenge remaining, *completely* glossed over by Fuller and presumably left as an exercise for a future generation of supervillains, is finding a material capable of handling the stresses of being part of such a colossal flying sphere.

Given the fact it needs to fly while also carrying a huge amount of weight, you need something with a colossal strength-to-weight ratio. That ratio can be hard to visualize, so instead we'll focus on an equivalent measure, called "breaking length." Breaking length is simply how long a vertical column of a material, supported only at its top, can be before it snaps—or "breaks"—under its own weight. A long breaking length indicates a material that's both strong and lightweight. Concrete has a breaking length of 0.44km. Stainless steel can get 6.4km long before it snaps. Aluminum can get 20km long, but we're going to consider two even stronger materials at the limit of both the breaking-length chart and humanity's means to produce them: Kevlar (breaking length: 256km) and carbon nanotubes (breaking length: over 4,700km, which is over a third of the diameter of the Earth).

Kevlar didn't exist when Fuller first proposed his floating cities (it was invented in 1964 by chemist Stephanie Kwolek), but it's a high-strength synthetic material used today for things like cut-resistant gloves and bulletproof vests. It's a recent but established material, and you can use it to

---

*At 829.8m tall, the Burj Khalifa is so tall that the sun can still be seen from the top floors after it has set at ground level, which impacts Muslims observing Ramadan. Dubai clerics have ruled that the fast—traditionally broken at sunset at ground level—must last two minutes longer for anyone above floor 80, and three minutes longer for those on the 150th floor or above.

cover your geodesic sphere in an airtight envelope. Kevlar costs about $32 a square meter at retail,* which means the 8,000,000-plus square meters of surface area on your geodesic sphere could cost around $256 million to seal. Not inexpensive, but seeing as that Kevlar also acts as a *bulletproof vest for your entire sphere*, it's probably worth the money. But that's just the covering, and we still need to build the structure itself.

Calculations show that aluminum could work (its weight means you may need to add helium to your sphere, or reduce air pressure inside to 0.9 atmospheres to achieve and maintain buoyancy), but there's a moonshot here if you're interested: the carbon nanotubes we mentioned before. They're another modern material unavailable to Fuller, but they just happen to have the largest breaking length of any material we've ever found or created. The catch is, they're still almost impossibly expensive, and the longest carbon nanotube created thus far was microscopic in width and just half a meter long, making this a material *almost*, but not *entirely*, relegated to the realm of science fiction. Most carbon nanotubes are sold as powder, an unorganized collection of fragments.

This suggests that if you really do want to build the strongest and lightest floating sphere it's likely possible for humanity to produce, you're going to have to first spend a fortune on research and development to make actual human-scale building with carbon nanotubes feasible. But there's a silver lining here: since there's already lots of research interest in this area, you won't be alone, and you have the advantage that it's pretty unassuming work. Most people don't hear "I'm funding research into stronger building materials" and think, "Ah yes, in order to build a colossal, sinister floating secret base far beyond the reach of all other humans and their petty 'laws' and 'morality'; this is the obvious endgame of any and all material sciences, say no more."

In conclusion, even without exotic building materials that are not yet

---

*Prices are extrapolated from the cost of a single 116m² sheet of Kevlar sourced in early 2021— you can definitely get a better deal if you buy in bulk!

and may never be viable for large-scale construction, a self-sufficient nuclear-powered Antarctic flying base will be difficult and hugely expensive and quite possibly unattainable by all save the richest and most visionary supervillains, but it's worth stressing this final point too: *it's not actually outside the realm of possibility.*

And spaces that seem like they should be impossible but somehow aren't are *exactly* where a supervillain operates.

A Black Mark Beside Your Name

Carbon nanotubes are used by Surrey NanoSystems to produce vantablack, one of the darkest materials on Earth: it absorbs 99.965% of all light that hits it. Perhaps something to consider when designing a sphere that you want to both capture heat from the sun and have a striking villainous aesthetic? (But keep in mind that a Surrey spokesperson, when considering the possibility of just a single full-sized bucket of vantablack paint, concluded, "I don't think there'd be much on the planet that would be more expensive.")

In 2016, any artistic use of vantablack paint was exclusively licensed to the studio of London-based artist Anish Kapoor, shutting out every other artist on Earth from using the material. This so annoyed fellow artist Stuart Semple that he eventually created his own shade of ultra-black acrylic paint with 99% absorption, which he then licensed to everyone on Earth—except Anish Kapoor.

POSSIBLE REPERCUSSIONS IF YOU'RE CAUGHT

I wouldn't worry too much about any minor laws broken in the construction of your sphere: pull this scheme off and—just like what happened with the Temples of Humankind in the sidebar earlier in this chapter—the sheer majesty of your floating base could induce authority figures to realize, in the face of your absolute accomplishment, that it's in their interests to align with you and grant you a pardon from any laws you may have technically broken to get to this point. And if not: well, jurisdictions were made to be floated out of.

But once you do that, it's still possible the nations you're above may suddenly realize they *do* want to police their high-up distant airspaces after all. Good thing for you that Antarctica is one of the few areas on Earth featuring unclaimed land (see Chapter 2: How to Start Your Own Country), which gives *you* legal standing to argue that the area above it is international airspace! Keep your sphere above Marie Byrd Land, a pie-slice-shaped piece of Antarctica running from the Antarctic coast to the South Pole, roughly at 158°W and 103°24′W, and you'll be floating above the largest unclaimed territory in the world.

But sadly, even when floating in international airspace above Antarctica, you're not fully out of the reach of the long arm of the law. The International Civil Aviation Organization (ICAO) is the United Nations unit that (as you may have guessed from the name) oversees civil aviation, and every country in the United Nations is a member—save for Liechtenstein, a European microstate with a population of fewer than 40,000 people and no international airport. And unfortunately, the ICAO has come up with several international agreements that (they believe) apply to you.

The first, published as part of the Convention on International Civil Aviation, is in effect even in international airspace. These "Rules of the Air" include regulations around things like instrumentation, lights, signals, cruising levels, rights of way, and parachutes (you're generally not allowed to make parachute descents unless it's an emergency: *disappointing*). It's all but certain you're going to completely violate many of these regulations, designed as they were for much smaller, non-sphere-based aircraft.

It gets worse once you look to the United Nations' Convention on the Law of the Sea (it also applies to aircraft; the UN is not the greatest at naming documents), which came into force in 1994. There are laws here that allow you to be boarded (Article 105, Seizure of a pirate ship or aircraft) and chased, if the chase starts in the chaser's territorial waters (Article 111, Right of hot pursuit). There are even rules against setting up your own radio or TV station (Article 109), so don't do that if you'd like to (figuratively) fly under the radar. But it's not all bad news: Article 87 subsection 1.d allows states to construct their own artificial islands, which is definitely a handy thing to keep in your back pocket. You are not a state,* but if you can find a country willing to let you fly under their flag, then you are simply a motivated private citizen taking advantage of the rights the UN guarantees to every nation in its membership.

Despite this disappointing international legal regime, remember this: laws can only be enforced if you allow someone to enforce them upon you, and existing *outside* the jurisdiction of any state is the whole reason you entered international airspace in the first place! If you can float high enough, fast enough, and free enough, then nobody will be able to tell you what you can and can't do in the privacy of your own 1.6km-wide flying sphere. After abandoning his seastead to the Thai authorities, Elwartowski wrote on Facebook that, for that short time he lived in his floating house, "I was free for a moment. Probably the freest person in the world. It was glorious."

It will be only glorious-er when experienced from a giant Kevlar-coated geodesic nuclear-powered sphere-city floating in the sky at the top of the world.†

---

*Yet. Again, see Chapter 2: How to Start Your Own Country.

†*Yes*, most maps put Antarctica at the bottom, but that doesn't stop it from being a completely arbitrary decision propagated out of inertia and a desire for consistency! If you can choose to

EXECUTIVE SUMMARY

INITIAL INVESTMENT	EXPECTED RETURN	ESTIMATED TIME UNTIL MATURITY

- **$4.4 billion** in aluminum rods (assuming a wholesale price of $2,900 per kilogram for the rods across a ballpark 1,500,000kg of aluminum structure)

- **$256 million** in Kevlar

- **$500 million** in Biosphere 2–style infrastructure

- **$700 million** in *Akademik Lomonosov*–style SMR nuclear generation and infrastructure

- **Total: $5.9 billion** in material costs alone, excluding construction and design

- **Limitless** (in terms of freedom)

- **Limitless** (in terms of a platform for supervillainy)

- **Limitless** (in terms of branding and marketing)

- **Limitless** (in terms of tourism potential)

- That said, in the interest of full disclosure: the base does not necessarily generate revenue, which means a loss of **$5.9 billion** is *possible*, though this could be mitigated to a mere **$2.95 billion** loss if you assume 50% of the value can be recovered upon the sale of a certified pre-owned secret base.

<20 years

---

build a floating sphere, then you can also choose to hang a map upside-down (or, dare I say . . . *right-side up?*)

HOW TO START YOUR OWN COUNTRY

> Lands doomed by nature . . . whose horrible and savage aspect I have
> no words to describe; such are the lands we have discovered, what
> may we expect those to be which lie more to the South, for we may
> reasonably suppose that we have seen the best as lying most to the
> North, whoever has resolution and perseverance to clear up this
> point by proceeding farther than I have done, I shall not envy him the
> honor of discovery but I will be bold to say that the world would not
> be benefitted by it.
>
> —*Captain James Cook (1773)*

Human children are born in a state of total helplessness: if they are to survive, those around them must feed and care for them, meeting their every demand until they are old enough to care for themselves. Thus, we move from a state of having our (generally fully selfish and self-interested) demands met as soon as they're understood (babyhood) to one where an increasing number of our desires are subservient to the demands of others (i.e., toddlerhood and beyond, wherein people with authority won't let us do everything we want because "it's bedtime" or "it's someone else's turn to play with that toy" or "oh lord that device might obliterate you and everyone else in a 10km radius, where did you even find that, *put that down this instant*").

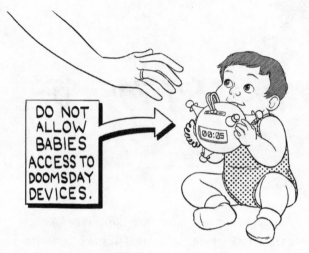

*This is not a parenting book, but this is nevertheless **explicitly excellent** child-rearing advice that you won't find in any other parenting book.*

Your fantasy of a world in which you don't have to answer to anyone, where what you say goes and your rule is law, and where you're surrounded by people who express their love for you and will also do anything you say—this is *functionally universal* to most of humanity. We all got a taste of it when we were born, and we see it still in the lives of dictators, regents, monarchs, and, to a lesser extent, presidents and prime ministers of democracies.

You get it by starting your own country.

BACKGROUND

Countries are sovereign areas of the Earth: they have the full and complete power to govern themselves and the land they occupy. This means the first step of starting your country is to get some land to build it on. Under current international law—itself based on criteria dating way back to Roman law—land and sovereignty can be acquired in one of four ways: you either find it, are given it, make it, or take it. Here's how each of those break down.

Finding Land

This happens when you discover an area of the Earth where nobody else lives and that no other nation knows exists, then claim it for yourself. Nation-building in this way was a popular pastime over a lot of history—especially during the era when Europeans decided that Indigenous people already living in the Americas somehow didn't preclude them from "discovering" and claiming their continents—but it's been stymied lately by the fact that humans and their meddling countries are *already* everywhere, leaving no undiscovered spaces remaining for an ambitious person to step out and form a new nation, conceived in liberty, and dedicated to the propositions of peace, order, and good villainy.

On a smaller scale, they ended in the 1960s with the invention of satellite imagery of Earth, a technology that quickly confirmed there were no more uninhabited islands left to find. On a larger scale, they've been over since before 10,000 BCE, which was the end of the glacial maximum of the last ice age, when sea levels were at their lowest and polar glaciers were at their biggest. It was during this glacial maximum period that humans migrated across what we now call the Beringia land bridge. Crossing a strip of land temporarily connecting Asia to North America near modern-day Alaska, these people found two continents of wilderness fully untouched by human hands, pristine in a way that's no longer possible on Earth. It was a new world filled with now-extinct megafauna like giant sloths, giant tortoises, giant beavers, and giant condors, mastodons, and mammoths, and even saber-toothed (and scimitar-toothed) cats—all of which began to be hunted and eaten to extinction by those migratory humans and their descendants. So, unfortunately, today this option is pretty much a nonstarter . . . failing, of course, the death of a large percentage of humanity.*

Megafauna's Megaextinctions

We don't know for *certain* if humans are the only culprit behind the extinctions discussed here, as there were climate pressures happening at the same time too. But we do know that throughout history, whenever humanity finds itself someplace new, extinctions tend to follow—especially the extinctions of large, easily hunted, *likely delicious* megafauna such as the American species previously mentioned, New Zealand's giant moa birds, and Australia's colossal Diprotodon, the largest marsupial to ever walk the Earth. Some recent evidence suggests humans may actually have been in the Americas as far back as 30,000 BCE, which is used to explain the megafauna extinctions occurring back then too. Extinctions due to human activity have been described as humanity's "most enduring legacy." (A supervillain, thinking of their own incredible legacy-in-progress, would hasten to add "so far.")

*Do not kill a large percentage of humanity. It gets *inconveniently* messy.

Being Given Land

Land is ceded when another nation more or less voluntarily gives it up to some other country—usually by that country purchasing it outright, by negotiating for it in a treaty, or by winning it through adjudication in the United Nations International Court of Justice.

Making Land

This occurs either through geologic activity (like when a volcano on land you already own erupts lava into the ocean, thereby producing more free real estate for you) or through the intentional acts of humans (like in Monaco, where land reclamation efforts have made one of the most expensive, wealthiest, *and* tiniest nations on Earth 20% larger—to 2.1km²—by expanding its shoreline out into the Mediterranean Sea).

Taking Land

There are three ways to take land from someone else:

1. You can get it through **stealth**, wherein you begin living and working inside a country as if it were your own, and hope the other country doesn't notice or object for so long that you can eventually claim the international equivalent of squatter's rights. This process, called "prescription," is rarely applied in the modern era, but it was used in 1953 when both France and the UK claimed fishing rights off some small, largely uninhabited islands and rocks known as the Minquiers and Écréhous groups, most of which are actually submerged at high tide. Both sides claimed ownership based on a different reading of the same medieval treaty from 1204. The aforementioned UN International Court of Justice decided the medieval document had less importance now than who had been actually exercising sovereignty on the islands since it had been signed, and awarded them to the UK.

2. You can get land through **force**—in other words, conquest. However, after two horrific world wars in the twentieth century, the international community decided there were maybe *some* downsides to running a planet this way, and such wars of aggression were criminalized and are in theory no longer recognized as a valid way to extend national borders. *Whatever.*

3. And finally, you can acquire land through **soft force**, wherein you recruit a sufficient number of the people living in a country to your cause, which, if successful, results in revolution and a new political system more favorable to your interests.

Now let's look at how others have tried, and failed, to exploit these possibilities.

THE INFERIOR PLANS OF LESSER MINDS

Finding Land

This is completely out for our purposes, since, as we already saw, all lands have already been discovered. You *could* attempt the path of Christopher McCandless, who, in 1992, finding himself in a fully explored world, tried to capture the mystery of an unexplored one by throwing away his map and walking off into the wild. Unfortunately, a few months later he starved to death alone in the Alaskan wilderness on land that someone else owned, so let's keep looking at your other options.

Being Ceded Land

Another nonstarter. It's unlikely any nation would seriously negotiate with you as an individual without you *already* having some other claim to the land and/or sufficient military might to credibly threaten to take said land by force—and if you have that might, you already have the strength of nations and can do what you want. The chances of orchestrating a purely financial transaction to procure land are similarly bleak: while nations have in the past purchased land for themselves—usually but not always from other nations*—there has to my knowledge not been a case in recorded history wherein a nation has sold land to an *individual* for the purposes of that individual setting up their own competing nation.

Of course, this hasn't stopped people from trying. In 1973, American fugitive financier Robert Vesco, wanted by the U.S. government on charges of securities fraud, escaped by fleeing on his corporate jet to countries without extradition treaties. While on the run he began negotiations with the island nation of Antigua and Barbuda to purchase Barbuda in order to found his own free and sovereign island state (and, presumably, to rename the nation to "Just Antigua Now"), but the government refused his proposition. (Vesco's story ends in 2007 when he died of lung cancer in Cuba while in the middle of serving a sentence for fraud, though friends suggested that him faking his own death and escaping in disguise would've been *classic* Vesco.)

Making Land

This only works if this new land shows up outside the borders and territorial waters of any other nation, otherwise it's automatically claimed by the nation it originated in. And if you do wait for a miraculous oceanic volcano to form and produce new land—or if you try to build it yourself through a huge amount of offshore terraforming—you still need to be lucky, fast, *and* able to defend your claim from any other nations who might want it for themselves.

In 1971, millionaire Michael Oliver and two other Americans hired barges filled with sand to turn the Minerva Reefs—the remains of an

---

*In 2012, Japan purchased a handful of uninhabited islands from the Kurihara family for ¥2.05 billion, but China—who already believed the islands were theirs—refuses to recognize what they called a violation of their territorial sovereignty. The islands are close to potential offshore oil and gas reserves, which is probably why both nations are so interested in shoring up claims to them.

ocean volcano so completely eroded that nothing above water still stands—into an artificial island and home for a new nation of his own design. Oliver named it the Republic of Minerva and intended to use initial investment in this new land (as well as the sale of official souvenir coinage) to fund more sand-filled barges, with the goal of eventually enlarging his republic to over 10km². On January 19, 1972, Oliver declared his republic's independence.

Within a month the neighboring states had met and agreed that Tonga, which, at around 400km away, was closest to the reef, had the only legitimate claim on this new island. By the nineteenth of June 1972, a Tongan expedition had chased away Oliver's people, raised their own flag on Minerva, played through their national anthem once, and, their mission accomplished, returned home. Left unattended and unmaintained, the fill seeped into the sea and the Minerva Reefs are again underwater.

So making land isn't an actionable plan either—though if you're handy with a boat, be sure to refer to Chapter 1 for some fallback options for floating cities.

Taking Land Through Stealth

This is generally no longer viable given satellites and other forms of pervasive global surveillance: you can't expect to hide out in another nation and not be discovered. What's worse, successful claims relying on prescription have always been between nations: it's deeply unlikely any nation (or any International Court of Justice) would recognize the claim of an individual, even under this criterion.

The closest this has come to happening in the modern era was probably by Alphonse Le Gastelois. In 1961, after being falsely accused of being a serial rapist and having his house burned down by angry neighbors, Le Gastelois left his home island of Jersey and began living on a smaller and otherwise-uninhabited island in the nearby Écréhous reefs. After a decade of living there alone, he formally requested that Queen Elizabeth II recognize his possession of the island as an independent entity under Norman law dating to 911 CE. Her Majesty refused. (The actual rapist on Jersey was caught in 1971, and in 1975, Le Gastelois returned to Jersey, where he remained until he died in 2012.)

Taking Land Through Force

Again, this process is no longer recognized by the international community and simply invites the force of larger and much more established nations to come down on you in response. Some have tried sidestepping the need to possess any real power by purchasing land, declaring their independence, and then simply hoping their parent state won't attack in response. Unfortunately, history suggests that even if you avoid armed conflict, few people around the world—and none of them in positions of power—will take you seriously.

An example here is the Principality of Hutt River, which began in 1970 when Leonard Casley, an Australian farmer upset over quotas that prevented him from selling all his grain, declared that his 75.9km² farm had seceded from Australia and crowned himself His Majesty Prince Leonard I of Hutt. Over the years the principality became a minor tourist attraction and eventually issued its own coinage and stamps, but it never achieved any international recognition or legitimacy. When his son Graeme—who inherited the kingdom after his father abdicated shortly before his death— saw tourism dry up during the COVID-19 pandemic in 2020, he shut down the kingdom after a reign of only a few years, selling it to the state of Australia to cover the over $2 million USD already owed in unpaid back taxes.

There's also the example of the Principality of Sealand, which began in 1967 when pirate radio broadcaster Paddy Roy Bates began squatting on an abandoned antiaircraft gun platform built by the British during World War II, in what were at the time international waters (updated laws now have Sealand firmly within British territory). Like Hutt River, Sealand also sells stamps and coins and offers itself as a tax haven, but it has even less hope of international recognition: the United Nations' Convention on the Law of the Sea states that artificial islands do not possess the status of real islands and therefore have no effect on sea territory rights. Sealand's actually done better than most: in 1968, after engineering professor Giorgio Rosa built his own 400m² tower in shallow waters of the Adriatic Sea just 11km off Italy's coast and declared it the independent Republic of Rose Island, Italian forces took it over and blew it up.

Taking Land Through Soft Force

This option is risky, because revolutions don't always go the way you want them to. Consider the case of Maximilien Robespierre: he came to power in 1792 during a period of the French Revolution known as the Reign of Terror after successfully arguing that the former king of France should be guillotined. Gaining power, Robespierre then helped write a new constitution (the French Constitution of 1793, never implemented), a new state religion (the Cult of the Supreme Being, never popular), and facilitated the guillotining of over 16,000 French citizens before being guillotined himself.

But "risky" doesn't mean "impossible," which makes soft force the most viable option we've got for acquiring land for a new country. Unfortunately, even this starts to fall apart upon inspection. Suppose you foment revolution, and you win, and you either take over an existing nation or carve off some revolutionary fraction of it for your own. Great!

... So now what?

Form a democracy, and you're becoming a mere politician, someone whose job it is to keep others happy to ensure your own survival and comfort. This is clearly not the absolute power we seek. You *could* start a monarchy and declare yourself king, queen, or regent, but let's eschew half measures and go all the way to declaring yourself a supreme dictator whose every whim is law. And yes, authoritarian regimes get a bad rap, but it's possible that's just an historical accident, right? You could easily imagine a dictator who's so enlightened, so clever, so wise and correct and perfect that their every decision only *helps* the nation and the people in it. And with a little villainous self-confidence, you could also imagine this person to be *you*. We have no evidence it's not, after all, and there's nothing that prevents a prudent and discerning dictator from making choices that are best for everyone, yes?

Unfortunately, there *is* something that stops dictators from being perfect rulers, and what's worse, it also impacts kings and queens and CEOs and prime ministers and presidents and democratically elected city council representatives and people in every other human-led power structure in the same way. It's human nature that trips us up here, or more specifically, it's *greed*. And this imposes some rules all rulers must obey.

Let's assume that you're already the perfect human who makes exclusively perfect decisions—and since you've already decided to read this book, *this seems like a fair assumption*. But even so, you don't exist in a vacuum, and you have supporters you rely on. In a democracy those are the members of the public (and of your political party) whose votes and support you need, and in a dictatorship they're a much smaller set of people whose acquiescence keeps you in power: military leaders, members of your palace court, personal guards, corrupt judges, etc. If you rely on *anyone* to keep you in power, then you need to keep those people happy, lest they stop supporting you and support a rival instead.

Welcome, incidentally, to the section where your supervillain book argues that all government is a form of bribery.

Because your supporters are humans and humans experience greed, you can keep them happy by continually giving them something they want. With a small set of supporters (like in a dictatorship), you can afford to offer them wealth, influence, security, and a subset of your power—stuff that nobody else gets to have. The problem with democracy is that there are too many voters to split your dictatorship-style favors across and still have them retain any value, so instead you offer them public goods: things that will make their lives easier, like roads, parks, health care, lower taxes, and so on. In all these cases, you give them *treasure*. And since even supervillains have finite resources, the loss of any treasure has costs and negatively impacts your ability to pursue the agenda you *really* want.

You may think, "Aha, but I am a clever supervillain, so I'll merely promise treasure to my supporters, and then once in power, I'll ignore them and pursue my enlightened dictatorial agenda instead," and you *can* do this . . . briefly. But remember: because of greed, there is always someone waiting in the wings who both wants the power you have and is willing to give your supporters more of what *they* want in order to get it. It's an attractive proposal, because the upstart wins more power, your supporters get more treasure, and the only loser is you. Once your supporters get unhappy, things like coups and assassinations (in dictatorships) and lost elections (in democracies) happen.

All of this leads to one conclusion: the only way to stay in power is to convince some set of other people that their lives are better with you leading them, which means your ambition must take second place to their greed. Heck, even *your* greed must take second place to their greed. It's harsh and disheartening, but the fact is that outside of the one-man-army armored-tank-suit of Doctor Doom that allows you to seize power, keep power, and enforce your will on others without relying on anyone else,[*]

---

[*]Before you get your hopes up, Doctor Doom's suit is both enchanted and powered by magic, and thus sadly beyond the scope of this book. It's too bad though. It's a flying suit that fires energy blasts out of its hands, and it has a cape *built in*.

you will *always* be beholden to others, always forced to subsume your will to their demands, and never truly be able to do whatever you want. There is actually no such thing as an absolute dictator—at least, not for long.

An unrelated gag comic after a very depressing series of paragraphs arguing for the inevitability of power corrupting even a perfect human being.

Thus we give up our fantasy of absolute control: it turns out we *do* live in a society, and because societies don't tolerate someone in power who makes the lives of *everyone* around them worse for long, your first job as ruler is always to keep at least enough other people happy that you get to keep being ruler. Console yourself with the knowledge that even with these limitations, power is still attractive, and people still crave it.

But if absolute control is out, and all the ways you have to acquire land for a new country are nonstarters, then what options remain for the sovereignty-minded supervillain here in the real world? Is there any way left for you to achieve even a *taste* of the control you get by having your own nation under you? It turns out, thankfully, that there absolutely is, and to learn how to exploit it, you need only turn your attention to the border of Egypt and Sudan, where you'll find a wedge of land called Bir Tawil: one of the few areas of *terra nullius*—land unclaimed by anyone else—left on Earth.

Bir Tawil

"But you already established mere paragraphs ago that there's no undiscovered land," you're saying, and that's correct. But while it's true that all land on Earth has been *discovered* . . . that doesn't necessarily mean that it's all been *claimed*. I apologize for perhaps misleading you on this matter over the past few pages, but some minor structural villainy can be justified when used to produce a memorable rhetorical flourish.

See? It says so right there.

Bir Tawil's origins date back to 1899 CE, when British forces—who were at that point occupying Egypt and had just recently conquered neighboring Sudan—drew a border between those two regions that looked, schematically, like this:

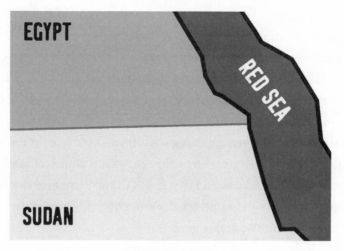

The Egypt-Sudan border?

But then in 1902 the British drew a new line of "administrative jurisdiction"—similar but somehow distinct from a national border—intended to reflect how the land was used by the people living there, which looked like this:

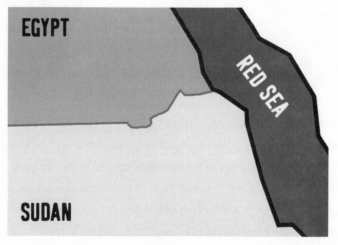

*Or maybe **this** is the Egypt-Sudan border?*

This new line was intended to move people considered culturally closer to Egypt or Sudan into the borders of those respective regions. Anyway, shortly after creating this mess, the British were induced to leave, and in 1956, what was (at least on paper) a single political entity broke into two when Sudan secured independence from Egypt. Perhaps inevitably, both Egypt and Sudan went with different editions of that British-drawn border, insisting that the version that gave *them* ownership over the Halayib Triangle—a larger, more resource-rich 20,580km² chunk of land with a seaport—was the valid one, and that it was the *other* country that was left with Bir Tawil: a less-desirable, landlocked, 2,060km² chunk of desert with no surface water and no arable land.

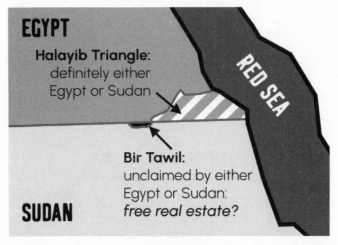

Your land borders?

Given that this dispute leaves both countries officially denying ownership over Bir Tawil in order to bolster their claims to the Halayib Triangle, several individuals have decided hey, if this land is unclaimed, then *I hereby officially claim it*. A selection of the countries these claims have produced include the Kingdom of Bir Tawil (official site: https://kingdomofbir tawil.blogspot.com, last update: 2010); the Kingdom of the State of Bir Tawil (official site: https://birtawilgov.weebly.com, last update: 2015); the Empire of Bir Tawil, whose official site was a now-deleted Facebook page; and the Kingdom of the Yellow Mountain, whose official site was a now-deleted Twitter account. But before you get too excited and set up your own governmental page on a free hosting service, you should know that claims like these are basically fantasy, ignoring the reality that:

1. There's more to nation-building than simply posting online, especially from a continent away, that you own Bir Tawil now.

2. Even if Egypt doesn't claim ownership of Bir Tawil and keeps it off all official maps, it still administers the land, and has done so since the early 1990s.

3. Just because both Egypt and Sudan claim the *other* nation owns Bir Tawil, it certainly doesn't mean that either is going to accept a third party waltzing in and claiming it, setting up a new country under their noses.

4. Most critically, all these claims ignore the fact that there are already people living nearby, including the Ababda and Bisharin tribes, who travel across Bir Tawil and even mine the land: while technically *unclaimed*, this land is neither unused nor uncontrolled.

Perhaps the most famous private claimant to Bir Tawil is American Jeremiah Heaton, who in 2014 decided to make his daughter's dream of being a real-life princess come true by starting his own kingdom. Unlike the examples we'd already seen, Heaton actually got the required paperwork from the Egyptian military authorities to at least enter the restricted area of Bir Tawil, and once there he planted a flag his children had designed, claiming the nation as "The Kingdom of North Sudan," with, of course, himself as king and his daughter as princess. Heaton's Facebook post gained attention in the media as a feel-good story about a local dad going the extra mile for his daughter, which led to it being optioned by the Walt

Disney Company for a feature film and to Heaton launching an Indiegogo crowdfunding campaign asking for $250,000 USD to help set up his new kingdom.

But a backlash began shortly thereafter, once critics noted how much Heaton's actions looked like imperialism and colonialism: this was, after all, the story of a white American, ignorant of the people on the ground, strolling onto the African continent and declaring that he now owned the place. Things soon got muddier when Heaton told a writer for *The Guardian*—who himself had visited Bir Tawil in 2010 and planted his own flag there, years before Heaton—that his Kingdom of North Sudan intended to offer itself to large corporations as a site that is both tax-, regulation-, and—as his country is an absolute monarchy—democracy-free. The Disney movie has not materialized, Heaton's fundraising campaign closed after having raised just $10,638, and comments on the campaign are peppered by backers disappointed that they've never received their promised nobility certificates. The kingdom's webpage remains online, where it speaks of the nation's ambitions to construct "the largest server farm ever built on the African continent" and promotes Neapcoin, the nation's already-defunct official cryptocurrency. Neither Heaton's nor any other claim to Bir Tawil has ever been internationally recognized.

But that's Bir Tawil, a land with its own people and complicated history. You're going to avoid the mistakes of Heaton and others by finding a place on Earth that's actually free of humans and their messy pasts. You're going to the largest legally unclaimed landmass in the world.

You're going to Antarctica.

YOUR PLAN

Let's examine the pros and cons of this frozen wasteland at the bottom of the world.

Cons:

- literally a frozen wasteland at the bottom of the world

- frozen (see first point)

- no arable land (see first point)

- hard to get to due to being at the bottom of the world (see first point)

- Captain Cook brutally dunked on it in the opening quote to this section

Now let's examine the pros:

- holds 90% of the world's ice and about 70% of its fresh water

- untold and untapped mineral resources (see sidebar on page 70)

- a place that, thanks to a warming climate (see Chapter 4), is becoming increasingly habitable over time

- even so, a still-challenging environment ensures populations will remain small in the near future, making it easier to maintain control of said population

- remoteness of continent ensures no national news organizations and their pesky reporters will be poking around

- Antarctica is the only continent without *any* full-time press, universities, military bases, judges, lawyers, *or* police departments

- also: ideal location for any villain with ice theming

If Antarctica's Mineral Resources Are Untold, How Do We Know They're There?

Two things tell us minerals are there: deduction and precedent. In 1974, the U.S. Geological Survey compared where Antarctica broke off from the supercontinent Gondwanaland back around 180,000,000 BCE and reasoned that the mineral resources located at the edges of at-the-time nearby Australian, African, and South American continents would also be found in Antarctica. Besides, they reasoned, "Large accumulations of minerals very probably occur in Antarctica, for no other continent is void of mineral deposits. The major question is whether these can be found and exploited economically."

So far, Antarctica has revealed gold, silver, iron, titanium, and coal deposits, evidence of diamonds, and potentially almost a quarter of the world's undiscovered oil and natural gas reserves. *Interesting.* Before a 1991 ban on development, the USSR, Germany, Argentina, and Chile all had their own projects to find mineral resources, and the United States had two uranium-search projects planned. *Even more interesting.*

And yes, since the early twentieth century there have been other countries with permanent bases already on the continent, and yes, these nations have used these bases to bolster their own competing land claims. But in 1959, after years of disputes and skirmishes—and with open conflict over these overlapping boundaries on the horizon—the twelve nations involved in Antarctica at the time* got together and agreed: decisive action was absolutely necessary to prevent all-out war. And so, after months of high-level negotiations and diplomacy, the nations finally emerged with a bold new agreement . . . to kick the can of worms named "Antarctic sovereignty" down the road so some other chumps would deal with it.

To be fair, the 1959 Antarctic Treaty did establish that the continent "shall be used for peaceful purposes only," bans military bases and nuclear testing from the continent, and "promote[s] international co-operation in scientific investigation in Antarctica"—but the treaty also explicitly states that nothing in it is a renunciation of "previously asserted rights of or claims to territorial sovereignty in Antarctica." No activities taken while the treaty is in force have any impact on existing or future land claims, no new claims can be made, and no previous claims can be enlarged. In other words, it freezes the continent exactly as it was on December 1, 1959, with all those overlapping and competing claims standing . . . only now with a shiny new international agreement in place to not worry about 'em for now.† It's actually the first arms control agreement of the Cold War and is generally regarded as a success. But this is how it leaves Antarctica:

---

*In alphabetical order: Argentina, Australia, Belgium, Chile, France, Japan, New Zealand, Norway, South Africa, the United Kingdom, the United States, and the USSR. (After the USSR fell, Russia assumed its Antarctic interests and responsibilities.)

†The United States and Russia never made any formal land claims, but they did reserve the right to make such claims in the future. In the interests of peace, both nations assured the international community that they *did* in fact claim land in the region; obviously they did, but the precise nature of those claims was a big ol' secret.

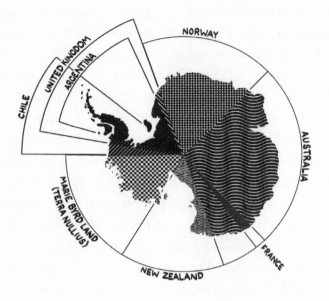

Land claims in Antarctica. Australia, who already owns the vast majority of a nearby continent, still claims almost 40% of Antarctica. At least everyone gets to share their own tiny sliver of the South Pole. Of note, and immediately discussed, is that giant wedge of terra nullius *representing about 15% of the continent!*

Marie Byrd Land is what you're interested in: that slice of unclaimed land we first met in Chapter 1 that, thanks to the Antarctic Treaty that's now been ratified by over 50 nations, is to *remain* unclaimed indefinitely. And what's more, your interests align with the Antarctic Treaty, because *you're not going to claim it either.* You're just going to set up a scientific research base there—permitted under international law—and then grow it until you've got, in effect, a small nation residing on some of this unclaimed land. Remember: you and your people are simply Antarctic researchers, obeying the same international treaties everyone else is, and not claiming *terra nullius* away from anyone.*

---

*You won't be the only one here: there are bases run by the United States and Russia in Marie Byrd Land too. But just keep your distance from them and you'll leave lots of room for your upcoming country. The unclaimed territory is over a million square kilometers big! There's plenty of room for everyone!

While you *could* simply proceed with the construction of the infrastructure your nation requires all by yourself, I recommend at this point that you secure the support of another nation. Do this by contacting one of the many existing nations interested in establishing a foothold in Antarctica and offering to set up your scientific base under their flag. *They* get a free science base and a seat at the table of other Antarctic nations, something available under treaty only to those nations with a serious scientific presence on the continent. *You* get the stealthy beginnings of your nation, the credibility that being part of an established nation provides, and an agreement to have as close to full autonomy as is possible within its boundaries. This concession shouldn't be too difficult to win, given the enforcement difficulties involved with you being so far from the mainland of every other nation on Earth. (Since Antarctica doesn't have its own legal regime, people there are subject to the laws of their home nation, which—of course—depends on that nation being willing and able to enforce them.)

Where Can You Find Food in Antarctica?

Since nothing in Antarctica grows outside of greenhouses and the wildlife is protected from hunting, you may want to try to get your sponsor nation to agree to bring food supplies once a year. If not (and if you don't have the self-sustaining hovering nuclear geodesic sphere of Chapter 1 to fall back on), you're going to have to pay to have food brought in yourself, but don't worry: it'll keep!

Food at McMurdo station—the largest research base on the continent—is brought in by boat once a year, then kept frozen until it's needed. Expiration dates on prepackaged food are routinely ignored, for two reasons: because the cold preserves things well, and because there's no other option.

As for what to bring: Laura Omdahl, who worked at McMurdo station for five years and ended up as store supervisor for all three U.S. bases, told *Nature* in 2020, "Doritos are the most sought-after treats down there. Everyone loves Doritos. So I had to put limits on them. Someone would want an entire case and I would say, 'I'm sorry, you can have two to three bags a day.'" If you're looking to recruit people already in Antarctica over to *your* research base, it couldn't hurt to make sure there's enough big ol' bags of Doritos for everyone on your supply boats.

Finally, when working out your food budget, keep in mind that people in Antarctica who work outdoors tend to burn more calories to stay warm, so aim for between 3,200 and 5,000 calories a day instead of the usual 2,000 or so and you'll have happy campers/citizens/henchpeople.

At this point you'll have invested a lot of time and money in establishing a research base in Antarctica: expect at least five years and $150 million USD, since that's how much it took to complete the Amundsen–Scott South Pole Station in 2003. While a research base certainly has its uses, it's still not actually a nation, nor has it made any claims to be. But don't worry: that actually means your plan is working perfectly. While you're building everything you need for your eventual nation/public research facility, you're also just *running out the clock.*

The clock in question is found in another Antarctic agreement: the Protocol on Environmental Protection to the Antarctic Treaty, a.k.a. the Madrid Protocol. The Madrid Protocol bans all mining on the continent, except in the case of scientific research, and it's been ratified by 34 nations (including all original 12 signatories to the Antarctic Treaty). However, it has a ticking time bomb embedded in it: a clause that allows it to be reviewed only after 50 years have passed since its signing, which lands in 2048. When that happens, it's fully possible that the Madrid Protocol won't be renewed.

While climate change *reduces* the viability of some regions on the planet (higher temperatures may make currently settled areas of the planet increasingly uninhabitable by 2050), it also generally *increases* the habitability of Antarctica, where warmer temperatures make things marginally more comfortable for humans, and melting ice makes it easier to reach the resource-rich land underneath. This could easily lead to nations no longer behaving so magnanimously regarding Antarctica, especially as any mining there would become more cost-effective and profitable. Once resource extraction begins, sovereignty becomes a critical question that can no longer be delayed, because you need to know whose land you're digging in—and because everyone will want to ensure that the land with the best minerals is the land *they* own. All this adds up to the real possibility that those land disputes that were frozen in 1959 might very quickly thaw out in the next 30 years or so.* But when that rolls around, if all goes according to plan, you'll have been occupying your land for so long that you and/or your adoptive nation have a chance at sovereignty through

---

*Dr. Doaa Abdel-Motaal, former chief of staff of the United Nations' International Fund for Agricultural Development and deputy chief of staff of the World Trade Organization, wrote a very detailed book exploring the nuts and bolts of this political scenario, called *Antarctica: The Battle for the Seventh Continent*, which—even though she doesn't explore the possibility of a *new* nation being founded there—was used as the basis of the arguments in this section. It's got a lot of useful sociopolitical details that could be exploited in the service of this villainous scheme, and since it never became a bestseller, it means there'll be fewer people also running this racket. You and I will just have to agree to keep it *our little secret.*

prescription, through earning a seat at the negotiation table with every other Antarctic nation as this international crisis unfolds, through other forms of diplomacy, or failing all that, through occupation or force.*

If Antarctic *terra nullius* collapses on or after 2048, and your sponsoring nation secures sovereignty over your land as it does, then congratulations: you are now functionally in charge of your own micronation and just one step away from securing actual independence. That final step, when the moment is right, is to negotiate for independence with your sponsoring nation, such that you are ceded your land and at last acquire the sovereignty you've long sought. This has been done successfully dozens of times in recent history—the collapse of the British Empire being just one salient example—and though it's often a slow process, it has been accomplished without armed conflict. You don't even necessarily need *full* independence to become a real country: Canada has functioned as its own fully sovereign nation with its own political and legal system since 1982, but with the British queen as its (largely symbolic) head of state, and with close political and monetary ties maintained with the United Kingdom. Britain gets to keep its illusion of empire and wacky tourist-attracting royal family; Canada gets its autonomy and constitution and independence, and *nobody has to die.*

---

*There's a chance your odds in an armed conflict in Antarctica might not be so bad: currently there are only a few thousand people in Antarctica at any given time, generally around 4,000 in the summer months and around 1,000 staying over winter. Prevent reinforcements from arriving—or staying—and you can keep it that way!

Canada's current supreme ruler, who, if she actually tried to exercise her powers, would immediately cause a constitutional crisis.

This is your best shot for founding and controlling a new nation on Earth as it stands today.* It's not guaranteed, and it'll take decades of work for even a *chance* at success, but it at least holds the possibility of leading to you both building, and controlling, your own Antarctic country. As for the rules for rulers we mentioned earlier: yes, you'll need to keep your supporters in Antarctica happy as discussed, but given the small population of your territory for the foreseeable future, that's simply equivalent to saying *you have to run a good team.* And if you don't want to do that, there is one legitimately evil (and not in the fun way) loophole in power structures: if the people beneath you can't physically overthrow you, then you can

---

*In my opinion. For a competing opinion, see Erwin S. Strauss's frankly bonkers 1984 book *How to Start Your Own Country: How You Can Profit from the Coming Decline of the Nation State*, in which he recommends using nuclear, biological, *and* chemical weapons against civilians in competing countries in order to secure your own sovereignty. Strauss notes with regret that "the details of obtaining and deploying weapons of mass destruction are beyond the scope of this book. They are covered in my book *Basement Nukes . . .*"

remain in power even if they're desperately unhappy. Antarctica is one of the few places on Earth where a properly positioned leader can ensure that they, and they alone, control access to food, information, and armaments, leaving unhappy people no other alternatives—unless they wish to take their chances being on their own in Antarctica's harsh wilderness.

THE DOWNSIDES

Like Bir Tawil, there are others who have claimed to own Marie Byrd Land without actually establishing a presence there. One example is the Protectorate of Westarctica, whose website allows you to become a non-resident citizen—all its citizens are non-resident—after completing a brief multiple-choice online quiz (sample question: "Where is Westarctica's claimed territory located?" Possible answers: "South America," "Africa," "The North Pole," and "Antarctica"*). Your claims will trump theirs, however, because unlike everyone else, *you'll actually live there.*

The usual downsides and caveats about living in Antarctica also apply: it's cold, it's lonely, in winter a single night lasts six months, people can get weird when isolated for a long time (see Chapter 1), and you're never going to be completely free if you have to rely on outside trade for food. But this can also be a benefit: it's the *frontier*. Antarctica already attracts tourists eager for just a taste of what the continent has to offer: if they knew it was an option, some of them might also want to escape the dreary mundanity and food security of modern life on a more permanent basis.

---

*According to the sixth edition of his combination autobiography/history of Westarctica, the ruler of the Protectorate of Westarctica, Grand Duke Travis, chose that title over king because "when you waltz around saying you're the King of Antarctica, you look like an idiot, and everyone knows it."

POSSIBLE REPERCUSSIONS IF YOU'RE CAUGHT

Listen: this is a long-term plan to win sovereignty from another nation, which means most nations, if they suspect what's going on, will be *immediately* hostile toward it. Sovereignty, like land ownership, is zero sum—for you to win, someone else has to lose. This has already been encountered by Westarctica's Travis McHenry when, in 2015, he got as far as securing permission from the Russian Antarctic Expedition to use their abandoned Russkaya research station as a Westarctica outpost. (He'd hoped to use it as part of a proposed reality television series documenting their attempt to visit, and then survive, on the continent.) When the United States got wind of these plans, the senior adviser for Antarctica for the U.S. Department of State contacted McHenry, telling him, "Failure to follow the applicable requirements could lead to civil or even criminal penalties," with McHenry stating afterward that "if I went to Antarctica without authorization from the U.S. Office of Polar Programs, I would be arrested upon my return to the United States."

On top of national hostility, there are international tensions too. Despite the signing of the Antarctic Treaty in 1959, other Antarctic nations have still tried to beef up their territorial claims by flying in pregnant women to their claimed parts of the continent, having them give birth there, immediately granting the child citizenship under the logic that they were born domestically, and then basking in the publicity that comes from the novelty of an Antarctic birth. Argentina did it first in 1978 when they flew in a then seven-months-pregnant Silvia Morella de Palma to their Antarctic base to give birth, which led to Emilio Marcos Palma becoming the first human born on Antarctica a few months later. In 1984, Chile retaliated—the Argentine claim overlaps with their own—by organizing the Antarctic counterbirth of one Juan Pablo Camacho. A handful of stunt births don't have a ton of bearing on national sovereignty, but when the stakes are this high, appearances matter and every little bit helps. It's also worth noting that since the 1959 signing of the Antarctic Treaty, nations have come together to peacefully build and operate an *international space station* in *literal space*, but we still haven't managed anything close to that here on Earth in Antarctica. The vast majority of Antarctic bases remain the property of a single nation, and while science is shared, stations are not: in 2011, of the 110 facilities in Antarctica, only two were joint stations, and those were split between just two nations. It appears the value of what's at stake—future sovereignty—is just too high.

And this means you'll want to keep your plans a secret, because if word gets out that you're setting up a new nation—even without evidence—then the game could be over before it begins. Remember, you're trying to heist both land and nationhood from several competing world super-powers *at the same time*, and even with your best efforts, the cold hard truth is that there's a chance that the almost three-decade-long plan described in this chapter won't actually work. There are no guarantees here, and I'm not going to lie to you and say there are.

But also remember this: even if you fail, there's still victory to be found in defeat! While you may not win control of your own nation at the South Pole, even deposed regents can escape to some safe-harbor country on a warm and isolated island, there to spend the rest of their days living in exile, working on their memoirs . . . and negotiating a deal for the rights to their life story with the Walt Disney Company.

EXECUTIVE SUMMARY

INITIAL INVESTMENT	EXPECTED RETURN	ESTIMATED TIME UNTIL MATURITY

$150 million for the construction of an Antarctic base, plus the ongoing cost of maintenance, supplies, and whatever elite-tier international negotiators are going for in 2048.

Untold billions, dependent on the market value of gold, silver, diamonds, oil, natural gas, uranium, land, and *freedom itself*.

<30 years, and the clock is ticking.

PART TWO

WHAT WE TALK ABOUT WHEN WE TALK ABOUT TAKING OVER THE WORLD

CLONING DINOSAURS, AND SOME PTERRIBLE NEWS FOR ALL WHO'D DARE OPPOSE YOU

> [J]ust as a picture is worth a thousand words . . . a living dinosaur would be worth a thousand court cases in the visceral effect it would have on schoolchildren.
>
> —Jack Horner (2009)

Every supervillain wants to make a good entrance. The best entrance it's possible to make is on the back of a dinosaur. Therefore, logic clearly dictates that every supervillain wants to produce, tame, and then ride on a dinosaur.

A successful supervillain who enters a room and everyone says, "What's their deal, and how did they become both so wildly attractive and so wildly successful?"

Your course is clear: you are going to bring dinosaurs back to life and *ride around on them*. This has the added benefit of playing God: a fun (that's why they call it "playing"!) and classic villainous pastime that every supervillain should try at least once.

BACKGROUND

Non-avian dinosaurs went extinct several million years ago shortly after an asteroid impact that occurred on what can fairly be described as "a pretty bad day" (see Chapter 4 for a brief timeline of how this went down). A few avian-type dinosaurs survived and eventually evolved into the birds you see around you today, and a few mammals survived and eventually evolved into the humans you see around you today too. Hello!

Before this happened, a vanishingly small percentage of these dinosaurs turned into fossils, usually by dying someplace very wet or very dry, getting rapidly buried by either sediment (in the wet places) or blowing sand (in the dry places) before scavengers could gobble them up, and then partially turning into stone over geologic time.*

And that brings us up to the present.

THE INFERIOR PLANS OF LESSER MINDS

One way to bring dinosaurs back to life would be through cloning, which means you'd need some dinosaur DNA: that double helix of linked biomolecules that makes up an animal's genetic code. In *Jurassic Park*, they got it from a mosquito trapped in amber.† The catch is that in real life, DNA breaks down too quickly for the amber idea to work—a recent estimate puts its half-life at 521 years, meaning that, *even under ideal conditions*, half of any dead dinosaur's DNA would be degraded and unusable in 521 years, and in 521 years, another half of the remaining bonds would be broken, and so on. Continue like that and all the DNA would be completely gone after around 6.8 million years, and there wouldn't be enough left to get useful data from in as little as 1.5 million years. Given that non-avian dinosaurs went extinct over 65 million years ago, the odds aren't looking good for dinosaur DNA recovery. In real life, we've never once recovered any DNA from anything that old, even ancient mosquitos trapped in amber: not from blood in its belly, not even from the mosquito's body itself. It's gone.

But let's put the DNA issues aside for a moment and focus on the other part of the problem: cloning an animal. In the past few decades we have cloned a bunch of beasts, ranging from sheep (Dolly, cloned from adult cells in 1996) to cattle (Gene, 1997), cats (CC, or "CopyCat," 2001), horses (Prometea, 2003), fruit flies (no names were reported for the five cloned flies, RIP

---

*We typically think of fossils as "bones that got turned into stone," but that's not quite the case. Minerals do strengthen some of the bone material, but most dinosaur bones contain much of the calcium they had while they were still alive.

†Both in the book and in the movie, at least. I haven't been on the water-based amusement park rides, but my research indicates they proceed from roughly the same premise.

fruit flies, 2004), and dogs (Snuppy, in 2005). But even in these successes there's more bad news for you: each of these animals was made through a technique called "somatic-cell nuclear transfer," which requires not just the animal's DNA but its *actual cells*: obviously a much more difficult trick to pull off with an extinct species.

In somatic-cell nuclear transfer, you take the nucleus (the DNA-containing core) of one cell and insert it into another one, basically replacing that cell's DNA with a copy taken from another animal. Finding an intact nucleus obviously poses a challenge for extinct species . . . but it may not be insurmountable, and the gastric-brooding frog is a fascinating example of this. These frogs were notable for how a female frog would swallow her own eggs after they'd been fertilized by a male, *outside* her body. These eggs would then hatch as tadpoles inside her stomach, and she'd vomit up live frogs later on. Unfortunately, the environment the gastric-brooding frogs lived in was damaged through human activity, and no gastric-brooding frogs have been seen anywhere since the 1980s. But in 2013, nuclei extracted from a single gastric-brooding frog specimen, frozen for 40 years, were combined with the egg cells of a still-living frog cousin, and this produced new living cells that—while they did divide—still failed to develop into a tadpole.

In fact, only a single extinct animal has ever been cloned: the Pyrenean ibex, a kind of wild goat you could find in the mountains of Spain. That is, until January 5, 2000, when the last known female, Celia, was struck and killed by a falling tree. And even here there were special circumstances: before she died, samples were taken of Celia's cells in an attempt to save her species, and afterward cloned embryos were created and implanted in surrogate mothers (closely related hybrids of other ibexes and goats— though scientists are working on developing artificial wombs for mammalian gestation, it's a long way off yet). Finally, one of the surrogates made it all the way through pregnancy without miscarrying, and on July 30, 2003, a new Pyrenean ibex was born by C-section. For the very first time in history, humans undid the extinction of a species.

It lasted less than eight minutes.

Despite all the efforts of scientists, the newborn ibex died shortly after birth, with an autopsy revealing that she was born with three lungs instead of two. The extra third lung—hard, like a piece of liver—took up all the space in her chest, meaning the other two couldn't fill with air. She never had a chance. After that, funding dried up for the Spanish scientists working on restoring their ibex, and no further attempts have been made, though Celia's cells have been cryonically preserved. In fact, today there are collections of cells from different animals—frozen zoos—throughout the world, some of them holding genetic material collected from species that have since gone extinct, all awaiting the day when we have both the technology and the willpower to restore them to life. (Some endangered-but-not-extinct animals have been cloned to introduce more genetic diversity to their small populations, as in the case of a black-footed ferret, which was cloned in 2021 from cells frozen in the mid-1980s.)

In any case, this technique to restore extinct animals to life may eventually be good news for the Pyrenean ibex, but it's still no good for you! You don't have any viable dinosaur cells, and without them—with just DNA and a dream—nobody on Earth has ever cloned an extinct animal.

Yet.

Here's the thing: some scientists *are* working on this problem, but instead of bringing back dinosaurs, they're focused on animals that went extinct more recently, like the passenger pigeon (extinction date: September 1, 1914, at 1:00 p.m., with some DNA still recoverable from the fleshy pads inside the claws of stuffed and preserved specimens in museums) and the woolly mammoth (extinction date: around 2000 BCE, and whose long-frozen bodies are sometimes found in Siberia). Both these efforts rely on taking the fragments of DNA that have been recovered, comparing them with a closely related species, and through that comparison trying to figure out where the DNA fragments fit, what they did, and which parts made the species unique. The idea is this: if you could identify the genes in mammoths that modern elephants don't have, if you could figure out which genes gave the mammoth its thick hairy coat, or its cold resistance, or its long tail, or made its ears smaller, or caused it to molt in summer, then all you'd need to do is alter the DNA of a closely related existing species so their genes match the extinct one, and presto: you've genetically engineered yourself a mammoth. That "all you need to do" there covers a lot of work in gene manipulation and gene insertion, but it is, in theory, possible. Some modified elephant cells have been produced in a lab with the cold-resistant hemoglobin that a mammoth would need, but nobody has produced any actual animals yet.*

The Passenger Pigeon's Last Flight

Remarkably, as few as 50 years before its extinction, the passenger pigeon was the world's most abundant species of bird—a position it had held for at least a hundred thousand years. They traveled in colossal, dense flocks of hundreds of millions of birds (possibly over a billion), which would darken the skies for days at a time, blotting out the sun and traveling in such numbers that the noise was described in 1871 as like "a thousand threshing machines running under full headway,

*Before the invention of genetic engineering, people had tried to "backbreed" extinct species into existence again: a process with the same goal, but a much rougher way of going about it. One of the most famous examples is that of the auroch, the species from which we originally domesticated cattle, which went extinct in 1627 CE. In the 1920s, the brothers Heck attempted to produce new aurochs by repeatedly breeding cows and bulls that they thought seemed the most auroch-like. This has produced some burlier cows (a breed now called Heck cattle), but they don't especially resemble aurochs—some other cattle, like the Spanish Fighting Bull, may actually be closer—and the process definitely hasn't resurrected any extinct species.

accompanied by as many steamboats groaning off steam, with an equal quota of [railroad] trains passing through covered bridges—imagine these massed into a single flock, and you possibly have a faint conception of the terrific roar." When they passed through an area, they'd consume all the fruits and nuts and other food in their path, and their droppings would fall like snow beneath them, leaving towns that "looked ghostly in the now-bright sunlight that illuminated a world plated with pigeon ejecta." We humans found that gross and annoying (and found their meat pretty tasty), so we killed them, and we kept on killing them until there were none left.

The last known wild passenger pigeon was shot by a boy with a BB gun on March 24, 1900, and the last specimen in a zoo to die was a female bird named Martha. Fourteen years after that last wild pigeon was shot, Martha died, alone in her cage, at the age of 29.

There are a few other pretty big problems in the way of bringing back dinosaurs: you don't have any prehistoric dinosaur DNA, nobody else has any prehistoric dinosaur DNA, there doesn't seem to be any hope of seeing even a *fragment* of prehistoric dinosaur DNA, and our best understanding says that all prehistoric dinosaur DNA disappeared from the entire universe more than 60 million years before our early ancestors even started walking on two legs.

For a lesser person, this is where the dream would die. But you're a supervillain. You think in ways that other people don't. What if you didn't *need* a full sample of prehistoric dinosaur DNA? What if you didn't even need a fragmentary one?

What if everything you needed was already here, waiting for someone to find it, lurking in the descendants of the dinosaurs that survived into the modern era?

YOUR PLAN

Evolution builds on what came before: it's an additive process. It's the reason many animals look so similar, and why we can group them into families. Humans have a brain at the top of a spinal cord, and so do dogs and cats and birds and fish and dinosaurs, and despite how different all these animals are, you can still see similarities between them: food goes in the front, waste comes out the back, limbs attach to the stiff rod of the spine, and so on. We're all vertebrates—a family of animals characterized by their spines—and if you go back far enough, we all have the exact same brain-on-top-of-spine-having, front-eating back-pooping ancestor. Each of these species just built on that—evolved from it—in different ways. (Not all animals are vertebrates, obviously, and non-vertebrate life-forms like arachnids and cephalopods all evolved from different ancestors with their own defining traits. Arachnids all have eight legs and an exoskeleton, and cephalopods all have large heads and a set of arms that evolved from the foot of a mollusk.) Even when an animal seems to have lost a trait—our earlier ancestors had tails, and we generally don't—that doesn't *necessarily* mean the genetic instructions for that trait were lost. They might just be suppressed. After all, when you were an embryo inside your mother's womb, you briefly grew a tail—only to reabsorb it a few weeks later, when other genes activated that stopped and reversed that process.

That's the process you're going to exploit. It's more than just genes that determine what an animal looks like; the timing and activation of those genes while the embryo gestates also matter. If you could interfere with the development of the embryo, if you could chemically suppress the gene that says "stop growing that tail" and instead let the process finish, you could produce a human with a tail. And the best part is, even though you will have made a new tailed human, you wouldn't have changed your subject's natural DNA! You simply intervened in how that DNA expresses itself during early development. You're not genetic engineering so much as you're merely building a *bespoke embryonic development environment*.

There are any number of extremely fraught moral and ethical concerns that people get tense about as soon as you hint you might be interested in messing around with some human embryos. But forget about all of them, because we're gonna deal with chickens! Chickens are one of the most studied and used lab animals, and we have been experimenting on them for over two thousand years. And it just so happens that chickens grow in eggs, which gives you, the science-minded supervillain, easy access to the complete embryonic environment as it develops!

Scientific Progress Goes "Cluck"

Those thousands of years of chicken experiments have yielded some great results! Back around 350 BCE, Aristotle opened chicken eggs at various stages of development to try to figure out what the purpose of the placenta was in humans. Later on, Louis Pasteur discovered (by accident) that chickens exposed to a weakened, slow-growing strain of chicken cholera could fight it off without dying, and then a few days later survive infection from a full-strength cholera strain that would kill other chickens: this was the first live vaccine produced in a lab. The chicken genome was the first bird genome sequenced, and we still produce influenza vaccines with the help of chicken eggs: influenza virus is injected into the eggs, which infects the embryo, which produces

Even better for you is the fact that birds are descendants of dinosaurs, which makes chickens some of their last living ancestors. If you look at dinosaur and chicken skeletons, scaling them so they're both the same size, it's remarkable how similar the two animals can be.

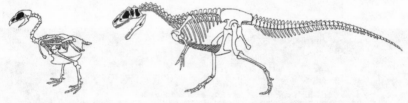

The skeleton on the right is from a dinosaur species named Allosaurus jimmadseni, *nicknamed simply "Allosaurus." The skeleton on the left is from a dinosaur species named* Gallus gallus domesticus, *nicknamed "the modern chicken, like from the farm or whatever."*

You can see the four main structural differences between a chicken and a dinosaur:

1. wings instead of arms and hands

2. a beak instead of a snout

3. no teeth instead of awesome teeth

4. a short, rounded butt instead of a distinctive long tail

In theory, *you could change this*. By timing the activation and suppression of genes at specific times, you could make a chicken grow a tail, and teeth, and a snout, and arms, and even hands and claws. The people you'll want to recruit are paleontologist Jack Horner and his assistants, who have been working on this project for decades. Horner believes we can make these changes, producing a chick that would hatch looking something like this:

All he needs is time and funding. He's not alone, of course, and if he's not available, many other scientists are studying (and influencing) the embryonic development of birds to see how different features develop. But Jack Horner is the one who has publicly said, for years, that he intends to combine these efforts to produce a dinosaur. He cowrote a book laying out his intentions in 2009, called *How to Build a Dinosaur*, but with the funding he's secured since then he has been focused mainly on convincing that rounded butt to become a theropod-like tail. Other scientists have focused on other parts: in 2015, a team led by Bhart-Anjan Bhullar suppressed the proteins that were tied with beak development and produced chickens with dinosaur-like snouts rather than beaks.* And some chickens have a

---

*His team didn't allow the chickens to hatch, though Bhullar believes they would've been able to survive "just fine." Another member of the team, Arkhat Abzhanov, said there "definitely

mutation that produces tiny conical alligator-like teeth in their beaks. This mutation is fatal and these chicks don't hatch, but it shows that with the right gene expression, the genetic information to grow teeth is still there. If you can get Horner's tail research to pay off, and then produce arms in a chicken instead of wings, and then put all the pieces together . . . well, there's your dinosaur. Though other scientists—including some of those who produced the snouted chicken embryos—are more skeptical, Horner says it could be possible within a decade or so. "When we succeed, and I have no doubt that we will, [it will be] sooner rather than later."*

So the question becomes: Is this a dinosaur?

And the honest answer is, no, not really. But here's the thing: even in *Jurassic Park*, a fictional world where anything can happen, *those* animals weren't really dinosaurs either. These imaginary scientists used frog DNA to fill in blanks in damaged DNA sequences, engineering animals that resembled what the general public *expected* dinosaurs looked like, but which were still genetically different. And if passenger pigeons or woolly mammoths are brought back, they won't be the same as the animals that came before either. The best mammoth we can hope for is a furry elephant that's adapted to cold, looks kinda like a mammoth, and fills the same ecological niche. The actual mammoths are all gone, and we don't know how to get them back.

---

would've been major ethical problems" with that. Lucky for us, supervillains tend to have different ideas of what major ethical problems look like!

*I found this quote in a *People* magazine article from 2015, however, so we're either closer than we think, or that opinion has been revised somewhat. In 2014, Horner told *The Washington Post* that when it came to expenses, "[We wouldn't need] more than $5 million. If we did have $5 million, then we would have three different labs working on it."

What If You *Did* Have Complete Extinct DNA?

Unfortunately, even that's not enough to fully restore the dinosaurs, mammoths, or passenger pigeons. There's more to many animals than their DNA, such as their *culture*. Many animals, including elephants, teach their offspring the same things their parents taught them when they were young: how to find or hunt food, what calls mean so they can communicate with others, and in the case of some (descendants of dinosaurs known as) birds, songs that have been passed down for generations. Even with perfectly cloned extinct animals, that knowledge and culture—and, likely, the environment that produced them—are long gone. When a species goes extinct, you can never truly bring back the animal that once lived, only a version of that animal. But make your peace with that, and you're in business!

Besides, you have the advantage that we know precious little about dinosaur cultures, so there won't be anyone who can complain when your dinosaur behaves differently.

Fundamentally—and genetically—your chickenasaurus is still a chicken. If it could breed with another chicken—which may be challenging, given how different those two animals would look (and possibly behave), the offspring would be perfectly normal chickens: you didn't change their DNA, just the development of an individual. There is no risk of the dinosaurs breaking out of the park and breeding, and if they do, there are several large and profitable enterprises in America dedicated toward the breading and frying of such animals. This is a plus, by the way, for the profit-minded supervillain: an inability to naturally breed more dinosaurs should both calm your critics and ensure that you can keep the monopoly on these creatures for at least a few years.

Finger-bitin' good.

Once you produce a chickenasaurus, you'll want to use what you learned on a larger animal: the ostrich. Ostriches are birds that are big enough to be ridden by humans and large enough to be a credible threat, especially when turned into the dinosaur-adjacent. Climbing on the back of an ostrich-asaurus is likely the closest you can get to riding a dinosaur, and honestly, when you storm through the halls of your enemies with a screeching, be-clawed, toothed beast that *looks* like a dinosaur, nobody is going to be testing its genetics to find out how similar your animal is to those that died 65 million years ago.

They'll be too busy fleeing in terror from a cackling demigod riding a dinosaur.

THE DOWNSIDES

Here's the big one: nobody knows if this will work. The genetic material for older features in an animal isn't *always* saved. Snakes—one of the odder vertebrates—are a great example of this. They don't have legs, but they used to have four of them—two at the top and two at the bottom. You can actually still see two tiny vestigial bumps on each side of a python's pelvis: the remnants of this lost feature. And boa constrictors still have the bones for tiny hind leg bones inside their bodies.

There's a gene that's necessary for limb development (along with organs and other tissues) called "sonic hedgehog," or SHH.* Snakes evolved the ability to suppress that gene from activating to grow legs, and while some snakes still have the DNA regulatory sequences required to turn on SHH, other snakes—like king cobras, which don't have any leg remnants—have lost large parts of that DNA. This isn't a surprise: since they have no effect on the organism, unused DNA sequences are much more susceptible to accumulating mutations or even being lost entirely. At some point in history there was a king cobra born with a degraded ability to activate the SHH needed to grow legs—and since it literally didn't matter and didn't harm the snake's fitness in any way, that mutation managed to spread throughout the population. To put it another way, if you had a library, you'd notice if I scribbled over or tore some pages out of a book you read often. But if I did it in a book you'd already decided to never read again, it'd be the perfect (though annoying and largely pointless) crime.

So there's no guarantee that all the features we want for a chickenasaurus are still there in chicken DNA. And if they are, there's no guarantee that these archaic genes will be so well preserved as to actually work: We can grow a chick with the structure of a dinosaur snout, but will it function

---

*Yes, this gene is named after the video game character that likes to go fast and collect rings.

well enough for the animal to breathe and eat? Are there other changes required in the skull to support having a snout instead of a beak that are missing? And even if there aren't and everything works, there's still no guarantee that the other changes required to make a chickenasaurus (the snout, the tail, the arms, the teeth) will fit together in one single chick in the right way, and even if *they* do, there's no guarantee that the resulting animal will function or even hatch: there could be brain structures or other features necessary for tails and hands that the chicken no longer has. We simply don't know.

But there's one way to find out . . . and to do some good science along the way. And it might just result in you making an incredible entrance both to the annals of science *and* your next dinner party.

POSSIBLE REPERCUSSIONS IF YOU'RE CAUGHT

Good news, American supervillains! In the United States, genetically modified organisms are regulated according to what they *do*, rather than what they *are*. Is your GMO a drug? Then you'll be dealing with the Food and Drug Administration. A pesticide? You'll have to talk to the Environmental Protection Agency. There's no federal legislation specific to GMOs. What's more, a 2014 report of the laws surrounding genetically modified organisms performed by the Library of Congress determined that "compared to other countries, regulation of GMOs in the U.S. is relatively favorable to their development." (Possibly because of that, the United States is already the world's leading producer of genetically modified crops, accounting for over 40% of the planet's total.)

If you've heard of it, you might be concerned about running afoul of the United Nations' Cartagena Protocol on Biosafety to the Convention on Biological Diversity, whose members pledge to contribute toward the safe transfer, handling, and use of any "living organism that possesses a novel combination of genetic material obtained through the use of modern bio-technology." (I know, I know: you didn't *technically* alter any DNA in producing your animal, just its developmental environment, but that will likely be a pretty thin distinction to someone facing down your birdasaurus). Signatories to the Protocol also pledge to use the precautionary principle when dealing with genetic modification: a look-before-you-leap philosophy that recommends caution, review, and study before jumping headfirst into new ideas or inventions that could be disastrous.

But the United States never signed it, so you're good there.

Universities have rules regulating animal testing and experimentation, but most supervillains are probably not operating in a university context, so you don't need to worry about these "respected institutions" and their precious "ethics boards." There are laws against animal cruelty in most countries (but not all: China, for example, doesn't yet have national laws forbidding the mistreatment of animals), but even as prospective supervillains, we'll stress that nobody is trying to be cruel here, and the goal is not to create an animal that will suffer. Quite the opposite: the goal is to hatch an animal that will be the most popular, famous, well-cared-for, and beloved animal *in all of human history.*

If you do find yourself attacked—either legally or in the court of public opinion—for artificially creating an animal that was not meant to be and which could now suffer because of that, remind your critics that humans have already bred millions of animals along those lines—animals that, because of their human-directed genetics, suffer more than their wild counterparts. Pugs, for example—an incredibly short-nosed breed of dog that exist because we humans made them and think they're cute—have compact airways that can cause them to have difficulties breathing, and their short noses have left them with few options in regulating their temperature. Dogs cool themselves through evaporation on their tongues, and a pug's stubby tongue leaves it susceptible to overheating and death from organ failure. But that horrible fate doesn't befall most pugs. Why? Because they're our *adorable pets* and we *love* them and will do *anything* to protect them and make them happy and comfortable. So too with your mighty dinosaur steed.

And remember: we are in the middle of a mass extinction *right now*, caused by human activity.* Habitat loss, hunting, pollution, climate change, human intrusions: the biodiversity on Earth has been decreasing in both plants and animals ever since we humans became the dominant species on the planet. All this to say: *you are not committing any crimes here.* Quite the opposite: you are doing humanity a favor, researching technologies that could be used not just to bring back something at least dinosaur-adjacent, but which will also produce all sorts of new discoveries and knowledge in both genetics and developmental contexts, and which could, one day, allow us to at least partially make amends for the species we've killed by bringing some version of them back.

Yes, there will always be people who complain that you're playing God, arguing that your work fails to address any urgent problem in our world, and that the millions of dollars this research would cost would be better spent preserving the species and habitats we still have, rather than trying to bring back old ones.

Their complaints will be difficult to hear over the roars of your dinosaur.

---

*It's actually the sixth great extinction in Earth's history—the fifth one was that asteroid 65 million years ago that took out most of the dinosaurs. Now it's our turn!

EXECUTIVE SUMMARY

INITIAL INVESTMENT	EXPECTED RETURN	ESTIMATED TIME UNTIL MATURITY
$5 million/year in lab and research costs for a chickenasaurus, more for an ostrichasaurus.	A 2018 *Forbes* examination of what Jurassic Park would earn in the real world produced an estimated $41.6 billion in revenue on $35.3 billion in operating expenses, for an annual profit of **$6.3 billion**.	**Approximately 10 years**

CONTROLLING THE WEATHER FOR THE PERFECT CRIME

Love is willing to become the villain so that the one who you love can stay a hero.

—Josephine Angelini (2015)

There are three parts to any crime: planning, execution, and escape.

People on both the demand and the supply side of crime (which is to say, cops and robbers) will tell you the same thing: that the perfect crime is one that's flawlessly planned, masterfully executed, and escaped from with consummate professionalism. They are wrong. The reason they're wrong is that they are laboring under a basic, entry-level definition of success and a basic, entry-level definition of perfection. You, as a supervillain, have evolved beyond such trivial limitations.

What's being forgotten here is that any crime you plan, execute, and escape from without a trace is by definition a crime whose author will never be known to either history or posterity. You are not becoming a supervillain to be *anonymous*. Please. Therefore, a stark conclusion presents itself: the perfect crime isn't the one you get away with.

The perfect crime is the one they thank you for.

What you need is a criminal act that will make you a hero not just to the people living on Earth today, but to the countless generations born in all our tomorrows. We're talking save-the-world-tier stick-ups here. Good thing I've got just the perfect caper in mind.

You, my friend, are going to *heist the climate*.

BACKGROUND

Climate change threatens not just people but entire countries and civilizations, and there's little point in taking over the world if there are no people left on it because they've all obliterated one another in a fight over livable land and drinking water.

You've definitely probably hopefully heard the origin and science of climate change before—it has to do with how carbon is used and goes back to the origins of life on this planet—but let's go over it quickly so we're all on the same page. I know your time is important, so I'm going to present it in the most efficient format there is for rapidly internalizing information: a fictionalized chat log.

2.7 billion years ago

x_cyanobacteria_x Registered User	i just evolved a new ability lol now i can turn light and carbon dioxide and water into fuel and release oxygen as waste i'ma call this . . . "oxygenic photosynthesis"
Other life Registered User	wait wtf OXYGEN?? ummmm that's poison to us!! ur gonna kill us!!!!!!!!
x_cyanobacteria_x Registered User	no lol see it's fine because iron oxidizes, and earth is covered in it!! so the iron will react with oxygen and rust and use it up that way. these and other "oxygen sinks" will keep atmospheric oxygen levels stable for uh i wanna say indefinitely??
Other life Registered User	oh cool thx

2.4 billion years ago

Other life Registered User	hey wtf!! millions and millions of years have passed and all the exposed iron on Earth has oxidized, so there's not enough non-rusted exposed iron left and now oxygen is building up in the atmosphere!! this "great oxygenation event" is going to cause mass extinctions!! UR GONNA KILL US

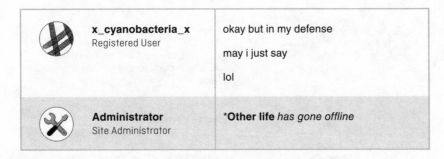

x_cyanobacteria_x Registered User	okay but in my defense may i just say lol
Administrator Site Administrator	*Other life *has gone offline*

2 billion years ago

Administrator Site Administrator	*New life *has come online*
New life Registered User	hey oxygen is way more reactive and we can use it to make all sorts of new chemicals and bodies, cool thx
x_cyanobacteria_x Registered User	lmao no worries

1.6 billion years ago

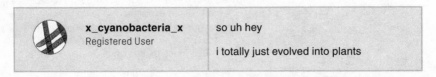

x_cyanobacteria_x Registered User	so uh hey i totally just evolved into plants

	Administrator Site Administrator	*__x_cyanobacteria_x__ has updated their avatar* *__x_cyanobacteria_x__ has changed their display name to __x_PLANTSBABY_x__*
	x_PLANTSBABY_x Registered User	haha sweeeeeeet gonna spread out of the ocean and colonize land now which means i will soon be found all over the entire planet, later haters

358 million years ago

	.oO amphibians Oo. Registered User	good news everyone we're the most common land animals, say hello to FROGS
	=-=ANTHROPODS=-= Registered User	UM EXCUSE ME!! WE'RE ALSO VERY COMMON ANIMALS WITH OUR EXOSKELETONS AND SEGMENTED BODIES!! SAY HELLO TO INSECTS AND SPIDERS AND CRUSTACEANS!!

x_PLANTSBABY_x Registered User	lol excuse u both, and forget animals because plants are still the most common form of LIFE and and AND we've just evolved WOOD and BARK and like wood and bark resist decomposition bacteria and funguses so effectively right now that when we die we just pile up forever and maybe in a few million years heat and pressure will transform us into such things as gas and oil and coal but for now it's just huge piles of dead plants everywhere especially in swamps maybe a few dead u guys in there too but we're the star of the show here so much so that i can only speculate that future generations will name this whole period after all the dead trees, perhaps "treetopia" or even "carboniferous era"
=-=ANTHROPODS=-= Registered User	WAIT SPEAKING OF CARBON IF YOU JUST PILE UP AND EVENTUALLY GET TURNED INTO FOSSIL FUELS THEN WHAT'S HAPPENING TO THE CARBON IN YOUR BODIES!!
x_PLANTSBABY_x Registered User	explain

	=-=ANTHROPODS=-= Registered User	TONS OF PLANTS BEING BURIED ON A MASSIVE WORLDWIDE SCALE WILL RESULT IN TONS OF CARBON—WHICH YOU TOOK OUT OF THE ATMOSPHERE AND USED TO BUILD YOUR BODIES—GETTING REMOVED FROM THE ECOSYSTEM FOR MILLIONS OF YEARS, POSSIBLY INDEFINITELY!!
	x_PLANTSBABY_x Registered User	i'm sure it's fine it's like, the climate's changed before and guess who was totally fine? i'll give u a hint this guy

240 million years ago

	Administrator Site Administrator	*DINOSAURS_RARR! has entered the chat*
	x_PLANTSBABY_x Registered User	hey welcome to the party, let's rule the planet together for around 175 million years and when u die let's make even MORE fossils and maybe even some more fossil fuels lmaoooooo
	DINOSAURS_RARR! Registered User	hi!!!!!!! I love it!!!!!!!!

66 million years ago

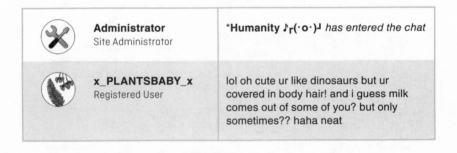	**Administrator** Site Administrator	*__GiAnT-aStErOiD__ has entered the chat at a speed of 32 kilometers a second and with the energy of 100 teratons of TNT_ *__DINOSAURS_RARR!__ has left the chat_
	x_PLANTSBABY_x Registered User	hey wtf?
	Administrator Site Administrator	*__GiAnT-aStErOiD__ has set the world on fire, blocked out the sun with dust and debris, injected sulfuric acid aerosols into the stratosphere (reducing sunlight hitting the planet by 50% and causing acid rain), and killed 75% of all life on the planet_
	x_PLANTSBABY_x Registered User	HEY WTF??????

200,000 years ago

	Administrator Site Administrator	*__Humanity ♪┌(·o·)┘__ has entered the chat_
	x_PLANTSBABY_x Registered User	lol oh cute ur like dinosaurs but ur covered in body hair! and i guess milk comes out of some of you? but only sometimes?? haha neat

10,500 BCE

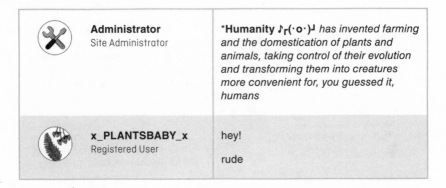 **Administrator** Site Administrator	*Humanity ♪ᵣ(·o·)ᴶ has invented farming and the domestication of plants and animals, taking control of their evolution and transforming them into creatures more convenient for, you guessed it, humans*
x_PLANTSBABY_x Registered User	hey! rude

1000 BCE

Humanity ♪ᵣ(·o·)ᴶ Registered User	Hello! I found some weird rocks that burn hotter than even the trees I cut down! I'ma call them "coal"!
x_PLANTSBABY_x Registered User	wow
Humanity ♪ᵣ(·o·)ᴶ Registered User	*Humanity ♪ᵣ(·o·)ᴶ has muted everyone else on this channel*
x_PLANTSBABY_x Registered User	W O W

1760 CE

Humanity ♪ᵣ(·o·)♩
Registered User

Hello! I just invented the Industrial Revolution. It's gonna be powered by burning fossil fuels such as coal and also oil and gas! We're gonna add so much carbon dioxide to the atmosphere! SO MUCH!

x_PLANTSBABY_x
Registered User

listen bud i know ur big into burning carbon and making carbon dioxide right now, but let me remind u that carbon ur burning has been sequestered underground for hundreds of millions of years

and as i'm sure u know carbon dioxide is transparent to light but reflects heat, so the gas acts like a greenhouse, letting light heat the surface of the planet but then keeping that heat from escaping into space

so uh maybe u want to think carefully before u release tons of carbon that's been buried for literally hundreds of millions of years, and then accidentally restore a different climate that existed before u even evolved

listen bro it's no bark off my back and i actually prefer a carbon-rich atmosphere, just giving u a heads up, bro to bro, bro

...

aaaaand i'm still on mute lol

1896 CE

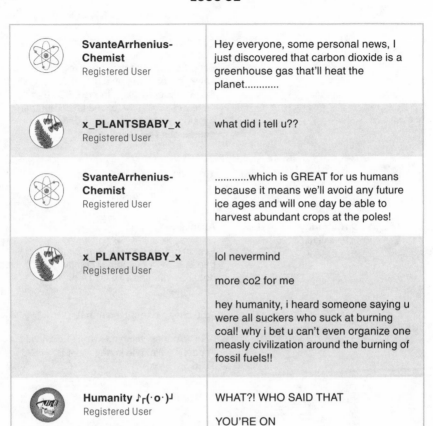	**SvanteArrhenius-Chemist** Registered User	Hey everyone, some personal news, I just discovered that carbon dioxide is a greenhouse gas that'll heat the planet............
	x_PLANTSBABY_x Registered User	what did i tell u??
	SvanteArrhenius-Chemist Registered User	............which is GREAT for us humans because it means we'll avoid any future ice ages and will one day be able to harvest abundant crops at the poles!
	x_PLANTSBABY_x Registered User	lol nevermind more co2 for me hey humanity, i heard someone saying u were all suckers who suck at burning coal! why i bet u can't even organize one measly civilization around the burning of fossil fuels!!
	Humanity ♪ᵣ(·o·)ᴶ Registered User	WHAT?! WHO SAID THAT YOU'RE ON

1988 CE

	Dr. James E. Hansen Registered User	Global warming has reached a level such that we can ascribe with a high degree of confidence a cause-and-effect relationship between the greenhouse effect and observed warming. It is already happening now.

	Humanity ♪ᵣ(·o·)ᴶ Registered User	OMG we need to fix this! Time for us humans to band together and solve this problem on a global scale, just like we recently did with the CFCs that were destroying the ozone layer, but then we stopped making them in only the course of a few years!
	Dr. James E. Hansen Registered User	Yes. Then it's agreed?
	Humanity ♪ᵣ(·o·)ᴶ Registered User	Absolutely!! (assuming we can still keep airlifting bananas in from another hemisphere because you know I GOTTA have my smoothies) (and cheap energy too haha) ANYWAY assuming this won't negatively impact my lifestyle in any way LET'S DO THIS

1997 CE

	Administrator Site Administrator	*Humanity ♪ᵣ(·o·)ᴶ has written and signed the Kyoto Protocol, pledging to reduce greenhouse gas emissions in order to slow, but not reverse, catastrophic human-induced climate change*

x_PLANTSBABY_x Registered User	aw excellent! u did it <3 i know we joke around here but real talk, domestication has been kinda sweet and it's real convenient to have human farmers and human gardeners and human houseplant owners looking out for yours truly through regular watering and seed propagation. i guess in a very real way our fates are now intertwined, and i always hoped u'd be good stewards of our small planet, the only island of life we know, this precious spaceship earth, a vulnerable and pale blue dot we share sailing all alone throughout the harsh and uncaring darkness of the cosmos
Humanity ♪г(·o·)┘ Registered User	It doesn't take effect until 2005, almost a decade from now!
x_PLANTSBABY_x Registered User	uh
Humanity ♪г(·o·)┘ Registered User	Also lots of us aren't gonna abide by it and the US won't ratify it and Canada's gonna quit it anyway and China isn't bound to reduce emissions at all and we're pretty sure they're going to become a major polluter in the next few decades because western nations already did everything they're about to do!!
x_PLANTSBABY_x Registered User	um
Humanity ♪г(·o·)┘ Registered User	We did it! We definitely did it!!
x_PLANTSBABY_x Registered User	oh oh no

Today

Humanity ♪┏(·o·)┛ Registered User		Wait the climate is still changing and its accelerating and we haven't solved it??
x_PLANTSBABY_x Registered User		listen the more u delay the harder it will be to change and the more change will be required to prevent catastrophic outcomes for ur species

a species that, i hasten to add, has gone from one billion ppl around 1804, to two billion in 1927, to on average a new billion ppl every seventeen years since then?? and just fyi because of that rapid population explosion, around five percent of all humans that have ever lived are alive right now, and if u don't do anything a significant portion of them are going to die as the areas they live in become inhospitable to human life, which makes this literally the highest-stakes problem u have ever faced

also u have thus far shown yourselves to be completely unable to effectively address a problem with shared and diffuse responsibility that requires a fundamental and expensive structural change (ur economies and electrical grids run on fossil fuels) but which also has a payoff that's decades away, whose distance has allowed u to kick the can down the road without any immediately visible consequences, and which, on top of all that, has proven hard for ur currently capitalistic species to make a profit from

without intervention some of the carbon dioxide u've already released into the atmosphere will remain for 100,000 yrs, longer than any time capsule or civilization or any structure created on earth by humanity so far, so unless u do something soon this might be the one thing that survives u all lmao

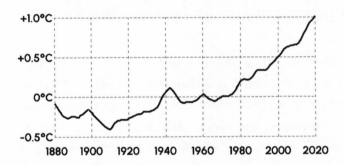 **Humanity ♪ᵣ(·o·)ᴶ** Registered User		Well I for one would sincerely love a fast, inexpensive, and global solution to climate change! whyyyyy is everything so hardddddddd
Administrator Site Administrator		*Humanity ♪ᵣ(·o·)ᴶ has changed their display name to* **Humanity ಠ_ಠ**

And that's where we're at today.*

Here is NASA data showing what Earth's (smoothed) temperature change looks like over recent history. This graph assumes that temperatures around the 1940s and '60s were "normal" (they weren't, but they're a useful comparison point!) and shows how much warmer or cooler years were compared to that baseline.

The average temperature change on Earth over the past 140 years.

---

*Svante Arrhenius—the Swedish chemist who first proposed that carbon dioxide could warm the planet back in 1896—did not conceal the fact that he was sick and tired of long, cold Nordic winters. His friend, researcher Nils Ekholm, went further and argued that the industrial coal-burning we were already doing was taking too long to warm up the Earth, and that humanity should speed things up by simply lighting on fire any and all coal seams found near the surface, thereby transforming the planet into a tropical paradise.

The average temperature on Earth has risen almost a degree centigrade since 1960. A degree centigrade may not sound like a lot, but each degree carries with it fundamental changes to our planet, from the obvious (the polar ice caps melting) to the more subtle (changes in where and when rain falls). There are people who will tell you this is nothing to worry about, that the Earth has changed temperature plenty of times before and will do so again. And technically, they're correct: the climate *has* changed a lot before! When the planet gets cooler, that's what ice ages are, and that Great Oxygenation Event 2.5 billion years ago did alter the planet's temperature while it was doing all those mass extinctions. But this misses the point that this climate change is rapid, unprecedented, human-generated, and potentially catastrophic. If you wake up in bed and see flames, you don't mutter, "Welp, houses have burned down plenty of times before and will do so again," before rolling over and going back to sleep. You leap out of bed, you yell for everyone else to wake up too, and you do what you can to fix it.

Here's what your options are.

THE INFERIOR PLANS OF LESSER MINDS

Climate change is unique in that it's a serious, life-threatening, and ongoing problem *that we already know how to solve.* If we didn't want melting ice caps and rising seas and drowned coastlines and mass extinctions and stronger hurricanes and greater droughts and temperatures rising in places to such heights that they're actually outside the range of human survivability, then all we needed to do was stop carbon emissions.

Done.

Only we didn't do it. Even today, we could at least *mitigate* the problem by reducing emissions. And yet, humans burned over 3 trillion cubic meters of gas in 2013—long after this problem was known. On top of that, we burned another 3 billion barrels of oil *each month* of that same year. And just to make sure our intentions were perfectly clear, we also spent that year burning a little less than 300 tons of coal *every second*. It all adds up to 30 billion tons of carbon dioxide released into the atmosphere in that year alone. Thanks to emissions like that, we've pushed the atmospheric carbon dioxide from an average of around 280 parts per million before the Industrial Revolution, to 310 parts per million in 1950, to over 410 parts per million today. Not only are we continuing to pollute, we're doing it faster than we ever have in history.

With all charity, these are not the actions of a species that's *especially* interested in fixing their carbon-emissions problem.

We're addicted to the cheapness and convenience of fossil fuels, and even after we've invented and invested in renewable fuel sources like solar, hydroelectric, and wind power, those are often used to supplement our ever-increasing desire for power, rather than to fully replace fossil-fuel generation. An obvious solution presents itself to even the neophyte supervillain: take over the world and use your iron fist to crush anyone who even thinks about emitting carbon. In fact, it probably seems like most of the problems you'll encounter in this book (or in real life) could be solved through what we'll call the T.O.T.A.L.W.A.R. maneuver: Take Over The Actual, Literal World And Rule With An Iron Fist.* Unfortunately, as we saw in Chapter 2, massive political reorganizations that put all of Earth's resources and activities under the enlightened rule of a single individual (you) aren't sustainable—and even if they were, they take time. At this point in the climate crisis, time is the one thing human civilization doesn't have. Besides, wars instigated by those foolish and petty enough to resist

---

*Technically, the rules of acronyms declare that this would not be the T.O.T.A.L.W.A.R. maneuver, but rather the T.O.T.A.L.W.A.R.W.A.I.F. maneuver. However, we're looking to sound imposing here, and a waif—defined as someone who appears thin, unhealthy, and weak—isn't the image you or any other supervillain should be looking to project. "Total war," in contrast, sounds totally badass, *especially* if you don't honestly interrogate how terrifying and dehumanizing any war, much less war of unlimited scope that disregards humanitarian interests and all international rules of conflict, truly is. So let's not!

you also have a nasty habit of releasing literal tons of carbon dioxide into the atmosphere: by the time you consolidate your power, the situation would be even worse. So what other options are there?

Well. It doesn't sound very villainous, but you *could* try cleaning up the mess.

So where do you start? The issue is that while the carbon dioxide added to the atmosphere since the Industrial Revolution is substantial, it's also diffuse, and nowhere near dense enough to easily find, capture, and remove. Carbon-capture machines—inventions that take carbon out of the atmosphere and put it somewhere in the ground (and are ideally powered by renewable, non-fossil energy)—face the challenge that even 410 parts per million is still just 0.04% of the atmosphere. They need to find the single molecule of carbon dioxide that's lost among the roiling sea of over 2,400 other molecules: tiny needles spread throughout a planet-sized haystack. Carbon-capture machines exist, but none of them can remove carbon efficiently enough and at a large-enough scale to solve our problem, nor are any likely to reach this scale in a useful time frame (see sidebar).

Die, Oxide!

A simple carbon-capture technology is to spray lye (sodium hydroxide) through the air: it reacts on contact with carbon dioxide to produce sodium carbonate, binding the carbon into a solid that could be buried. The challenge is that even our record-high 410 parts per million of carbon dioxide in the air means you're spraying an awful lot of lye to encounter a very small amount of carbon dioxide, and spraying lye takes energy—which, in most systems, produces more carbon dioxide. Similar technologies are used to scrub carbon dioxide where it *is* found in higher concentrations—for example, inside the smokestack of a coal power plant—but they only reduce *new* carbon dioxide being emitted and do nothing to solve the problem of the carbon dioxide that's already in the air.

If you're committed to this scheme, know that it's not fully impossible: one calculation showed that if you had carbon-capture technology running off nuclear power (no carbon emissions), you could actually remove all the excess carbon added to the atmosphere since the Industrial Revolution. All you need is 4,688 nuclear reactors (ten times more than there are active nuclear power reactors in the world today, and which, if they were all in the U.S., works out to a reactor every 2,000km^2) all operating constantly at full capacity (800TWh/year). Once you've got those squared away, the operating costs of the carbon capture itself will run you about $13.75 trillion per year, which over the 80 years it would take works out to *over a quadrillion dollars*. On the plus side: pull this off and you'd have control of 4,688 nuclear reactors, which has to have *some* upsides.

What about trees? They're the original carbon-capture technology. Could we just plant giant forests to store the carbon? Well, we could've . . . several hundred years ago. The amount of carbon dioxide in the air now means that even under ideal conditions capturing it would require over 9 million square kilometers of trees—that's over two Indias' worth of surface area, and about 40% of Earth's arable land. You can't cut your planetary farmland almost in half and still keep all 7 billion people depending on it alive. And starving billions of humans to death isn't supervillainy, it's just garden-variety evil. *You're better than that.*

So if we can't clean up the mess we've made, and we won't stop making this mess worse by burning fossil fuels, and even taking over the world won't solve things, then what's left?

I'll tell you what's left.

You, acting alone, are going to take control of the entire world's climate—by producing a powerful and sustained *artificial volcano*.

YOUR PLAN

When large volcanic eruptions go off, they can affect the climate. The massive 1815 eruption of Mount Tambora in Indonesia—the largest eruption on the planet in the preceding 1,300 years—caused a global temperature drop of 0.7°C and resulted in people in the Northern Hemisphere nicknaming 1816 "The Year Without a Summer," "Poverty Year," and "Eighteen Hundred and Froze to Death." One of the obvious ways volcanoes cool the climate is through ash: these tiny pieces of rock and dust physically block out the sun, cooling everyone and everything beneath their dark clouds. Unfortunately (for our purposes), the effect of ash is generally short-lived: the larger particles fall back to Earth quickly, and even the smaller ones get pushed down with rain. However, the very smallest particles can sometimes make it out of our troposphere (the more turbulent area of the atmosphere, where odds are you've spent 100% of your life) and up into the stratosphere (a higher, calmer section of the atmosphere above it), where they remain for months. You might think injecting ash into the stratosphere is the solution here, but you can do even better than ash. Erupting volcanoes also release something else into the stratosphere that's even more effective at altering climate: *sulfur dioxide.*

What the Atmosphere Looks Like

The atmosphere is divided into layers: the troposphere runs from where we live at the surface and goes up about 12km high. It's where most weather happens, where most water vapor and clouds are, and where the vast majority of airplanes fly. Above that is the much calmer and more stable stratosphere, which goes up to 50km above the ground. The air here has roughly a thousandth of the air pressure at sea level. Above that is the mesosphere, ranging from 50 to 80km, and sometimes called "the edge of space." This is where most meteors burn up on entry into our atmosphere. Above *that* is the thermosphere, ranging from 80 to 700km, where the atmosphere is so thin that each molecule in it travels on average a kilometer before it encounters any other molecules. This is where the International Space Station orbits, and if you enter it without a spacesuit on, you'll die. Finally, above that is the exosphere, ranging from 700 to 10,000km. Here there are gasses that are still under the influence of Earth's gravity, but they're so diffuse that they no longer behave like a gas and instead look a lot like an assortment of molecules that are constantly being lost to space. Above the exosphere, what's left of our atmosphere merges with the solar wind, and we call that "outer space."

When sulfur dioxide reaches the stratosphere, it combines with water to form sulfuric acid, which condenses into tiny droplets around particles of dust and forms a whitish haze. This haze reflects some of the sun's light back into space *before* it can warm the planet, reducing the light we receive on Earth and acting like a giant lampshade around the planet. It doesn't take much: reflecting just 2% of the light that would otherwise hit the planet would be enough to bring temperatures back to their pre–Industrial Revolution levels. What's more, this haze of aerosols (an aerosol is just a suspension of tiny particles of liquids or solids) can remain in the stratosphere for over a year before it eventually reenters the troposphere and returns harmlessly* to Earth. As you might recall from our Scientifically Accurate Chat Log, when that asteroid hit 66 million years ago, it also released sulfuric aerosols into the stratosphere, and the cooling that resulted helped kill off most survivors of the initial impact.

So you'll just have to be more careful than that.

Here's what you need to do:

1. Collect some sort of chemical precursor: we're going to use sulfur dioxide, but there are lots of other options (hydrogen sulfide, carbonyl sulfide, ammonium sulfate) if you want to get creative—anything that can produce a white haze will work.

2. Get your chemicals up to the stratosphere.

3. Spray them out in tiny droplets so they form an aerosol.

4. Repeat this process at least once a year.

5. You're done, there is no step five. You have successfully heisted the warming climate and replaced it with a cooler one.

---

*Well, almost harmlessly. Turns out there are *some* downsides to sulfuric acid falling from the skies? We'll get to it!

Besides being simple, this plan is *cheap*. Sulfur dioxide is easy to get: you produce it by burning sulfur (it's the only thing that the reaction of sulfur and oxygen produces), and sulfur is *abundant* on Earth. It's the tenth-most-common element in the universe, the fifth-most-common element on Earth, and the fourth-most-mined element by weight, so we're good here.

Getting your sulfur into the stratosphere is also quite affordable. While flying 20km above ground into the stratosphere is outside the range of commercial passenger aircraft (they tend to top out at around 14km up), we've already invented surveillance aircraft like the Lockheed U-2 that cruise 21km above ground, well into the lower reaches of the stratosphere. Buying a few dozen such aircraft and altering them to make them better suited to gas dispersion (larger cargo bays, an even higher operating ceiling) would probably cost around $7 billion USD. If you don't have your pilot's license, a balloon with a hose leading back down to Earth has been proposed, or giant airships, or you could even use large naval-style guns firing shells filled with sulfur dioxide that burst in the stratosphere (a technology that has existed since the 1940s). So far, so good: all the technology required here already exists, or could be made to exist relatively easily.

Above: a committed environmentalist.

Once you have the aircraft, operational costs come in at around $2 billion USD a year, and this includes the tens of thousands of flights necessary each year to create and maintain the sulfur haze. While those flights would be detected by the powers that be, there are airfields in every industrialized nation around the world. This makes it functionally impossible for a single authority to stop you or your machinations, especially if you launch your initial flights in a coordinated blitz.

How *Else* Can You Get Sulfur Dioxide
into the Stratosphere?

While aircraft are the most established and reliable technology for our purposes, they can be a little staid, and there's always something to be said for style. Rockets are always fashionable and could be designed to vent sulfur dioxide as they cut through the stratosphere on their way to orbit. This, while more expensive than airplanes, does result in tons of orbital rockets that you can then use for *whatever your heart desires*.

If shelling is more attractive, you may be interested in the British 13.5-inch Mark V naval gun, which was invented in 1912 and installed on the excellently named "super-dreadnaughts" of the Orion class. One such gun, with the less-excellent-but-still-passable nickname of "Bruce," was first used for stratospheric shelling on March 30, 1943: it fired up smoke shells 29km so that their dispersal patterns could be studied. If you want even more power, there's always Project HARP, a joint U.S.-Canada project whose goal was to use a colossal gun to *blast satellites into space*. Project HARP produced a supergun with a barrel 36m long that—before the project was shut down in 1967—succeeded in launching a projectile a full 181km above the Earth.

While actual scientists have produced actual scientific papers proving that *relentlessly shelling the very skies themselves* will never be the least expensive option to get aerosol

$2 billion a year plus $7 billion up front? In terms of evil schemes, this is child's play. For comparison, Elon Musk's net worth at the start of 2021 was $203 billion, which means, acting alone, he could both begin this project and fund it for almost a century, and he's not even a supervillain.* He's just *some guy*, among the dozens of nations, with both the budget and expertise to launch such programs. And when compared to projected costs of climate change, estimated at $600 *trillion USD*, this is a stone-cold bargain.

So if this project is so simple and affordable that one person could launch it, and so effective that the fate of our civilization could depend on it, why hasn't anyone done it yet?

THE DOWNSIDES

The obvious one is that this solution to climate change only lasts for as long as you keep seeding the skies: stop cold turkey, and your white haze will quickly fade, its aerosols will fall back to Earth, and the planet will return to whatever temperatures it would otherwise be at in the space of a year or two. And yes, years or even decades of warming all happening at once could be, to put it simply, a challenging survival scenario. But humans have adopted plenty of other technologies that, if we were to suddenly abandon them, would also cause massive death on a global scale! If

---

*To be fair, Musk *has* used some of his obscene wealth to launch a car into orbit via his private spaceship, which is at least supervillain adjacent.

we suddenly stopped fertilizing plants with nitrogen, for example, within the space of a year we wouldn't be able to make enough food for everyone alive today. We've long ago decided that using technology to keep everyone alive was worth the risk.

Obviously, you could use the dangers inherent in abandoning this plot to run blackmail on a global scale: "Pay me to keep doing my quirky fun geoengineering project, or I'll stop, and then *everybody's going to die*." But the beauty here is that you don't actually have to blackmail anyone! Once you bring this process to maturity, humanity itself now has huge motivation not only to *not* stop you, but to help you continue it, and even to take over doing it themselves. Any heist that results in the so-called "victims" helping you continue it indefinitely seems like one that is, objectively, pretty well designed.

That's another downside nicely sidestepped, but there's still the issue of stratospheric sulfuric acid returning to the surface as acid rain. And yes, there may be localized areas of extreme damage in places that, due to changing and unpredictable weather patterns, receive more than their fair share of acid rain than anyone else. But it's also true that humans would've died in *natural* climate-based deaths had you done nothing, and besides, let's not be so quick to dismiss how your white haze has turned the skies from their overly-familiar-if-not-slightly-tedious shade of blue to something much whiter! And have your critics not *also* noticed how your airborne particles scatter light to produce some truly beautiful and dramatic sunsets of a kind not seen since the disastrous eruption of Mount Tambora in 1815?

But honestly, whether you change the global climate with something as bold as sulfuric acid or something as staid as orbital mirrors (see sidebar), these are both distractions from the core issue, which is that of *culpability*.

What About Giant Space Mirrors?

It's true: you could avoid the downsides of sulfur dioxide by instead reflecting sunlight back into space via a planetary network of orbiting colossal mirrors! In fact, some critics might say that by not doing so, you're being greedy by choosing the cheaper sulfur dioxide technique over the more expensive but safer "giant mirrors in space" one. Remind these critics that a properly arranged network of orbital mirrors could also be used as a very seductive "ants under a magnifying glass" doomsday weapon, so if anything, by filling the sky with sulfuric acid *you're actually doing them a favor.*

When hurricanes strike, or lands flood, or drought arrives, we call those "acts of God" because we believe no humans could possibly be blamed for them. But adjusting the climate also adjusts weather patterns in ways that are difficult to predict, so now when hurricanes strike, or lands flood, or drought arrives, these will be taking place, in part, because of what you've done. There will no longer be acts of God, only acts of You.*

There is really only one way to mitigate your responsibility here, and that's by limiting the haze you produce. Some research suggests there may be concentrations of haze that would not push precipitation outside average ranges in more than a handful of regions: holding yourself to just 50%

---

*Please note that thou hath become God only as far as the weather is concerned, and only then when considering the limits of liability within certain legal jurisdictions. But still: not a bad outcome for a heist you read in a popular science book.

reduction in global warming might significantly change water availability in a mere 1.3% of land area worldwide. But this brings its own challenges: even ignoring those who live in that 1.3% of land area, the people in countries still suffering the effects of climate change that you've only partially mitigated—but not wholly removed—could blame you for not doing *more* to save them.

It gets real messy real quick, and that's before considering the complications that arise from those who for some reason believe that if there is to be a global thermostat, a lone supervillain should not be the only one controlling it. But remember: you didn't get into the supervillain business because you wanted to be popular or because you wanted to avoid encounters with international courts of justice. You got into the supervillain business because you read a very compelling science book that made you realize you could, *and probably should*, take over the world.

Besides the thorny issue of culpability, the global reluctance to put this scheme into action also stems from another, even simpler objection: it might not work. What we're discussing makes sense on paper, and it works when volcanoes do it, but we've never actually tried to put sulfates into the stratosphere at this scale before. Many scientists say more research is needed, while others say that the very idea proposed here is "untested and untestable, and dangerous beyond belief." And they may be right. We humans don't have the greatest track record at foreseeing *all* the consequences of our actions, whether at the level of the individual (see the Streisand effect), or of society (see American prohibition in the 1920s, our history of introducing species to new habitats, or the Scunthorpe problem*). There's a very real chance that we *shouldn't* run global experiments

---

*In an effort to prevent you from putting down *this* compelling book to go look these up somewhere else, the Streisand effect began when famous singer-actor-director Barbra Streisand tried through legal means to have a photo of her house that just six people had seen removed from the internet, resulting in drawing the attention of hundreds of thousands of people to it *and* getting a whole effect named after her; American prohibition of alcohol didn't stop alcohol consumption but instead pushed it underground and supported the rise of organized crime; our history of introducing species to new habitats includes many instances of this leading to the

that could negatively impact the habitability of the only planet we've got. (On the other hand, there's also a chance that by simply putting these theories into practice, you're not just becoming a supervillain: you're becoming a scientist performing one of the greatest practical research experiments this planet has ever known, so—call it a toss-up?)

"Untested and Untestable, and Dangerous Beyond Belief"

This quote comes from James Rodger Fleming in his 2010 book *Fixing the Sky: The Checkered History of Weather and Climate Control*, though, to be fair, he's considering not just our scheme here, but also many other arguably zanier schemes, including "flooding the ocean with white plastic so more sunlight is reflected," "brightening the clouds so that they're whiter and more sunlight is reflected," and "fertilizing the ocean, causing blooms of carbon-dioxide-absorbing phytoplankton in the hope that some of that algae will sink to the bottom of the ocean and trap the carbon there." Those inferior schemes don't even involve a single rocket, giant gun, *or* worldwide fleet of specially modified stratospheric aircraft!

extinction or near-extinction of other unintended species; and the Scunthorpe problem refers to computer systems designed to filter out bad words accidentally censoring more innocent speech, which has affected everyone from people named "Weinersmith," to those curious about Super Bowl XXX, to those lucky enough to have graduated magna cum laude.

Look, I'm not going to tell you that this is a perfect solution. Even with ideal results, this scheme only addresses the "warming" part of climate change and doesn't actually remove any carbon dioxide from the atmosphere. Atmospheric carbon still has serious effects on the planet beyond just temperatures: for example, the oceans acidify as they absorb more carbon dioxide, which causes things like the bleaching of coral reefs. This scheme involves humanity crossing a line it's never been willing to cross before: the intentional manipulation of the global climate. It's not the end of nature, but it is the end of one conception of the natural world. All of this is true.

But it could also buy the world breathing room: time to invent some new miraculous carbon-capture technology, time to wean ourselves off fossil fuels, time to change our consumption habits or invent cold fusion or do any number of other things we could be doing to save the world we share. Assuming it doesn't kill us all, it could postpone many of the worst parts of any climate apocalypse for decades, during which an enlightened, self-motivated leader—*hmm, I wonder if any of those could be reading this book right now*—could help direct the planet toward a more sustainable, less disastrous future.

Like I said, it's not the perfect solution.

But it could very well be the perfect crime.

POSSIBLE REPERCUSSIONS IF YOU'RE CAUGHT

You are truly entering uncharted territory here: a heist of this magnitude, something that both alters the planet's climate and affects every living thing on Earth has never been achieved before. The closest thing to a law governing this scenario is probably the United Nations' Environmental Modification Convention, also known as the Convention on the Prohibition of Military or Any Other Hostile Use of Environmental Modification Techniques.[*] The word "hostile" is the important one there: this is a treaty banning weather warfare, and you're not doing this to start wars—you're doing this to save the entire world, which by any definition is just about as far from "hostile" as you can get. You're further protected by the fact that the UN Security Council has no authority over individuals, only member states, and besides: Article III of the convention states, "The provisions of this Convention shall not hinder the use of environmental modification techniques for peaceful purposes," so you should be good.

There's also the United Nations' London Convention on the Prevention of Marine Pollution by Dumping of Wastes and Other Matter. It could conceivably be argued that when your haze falls back to Earth you're dumping waste at least partially in the ocean, but again, you don't have much to worry about here: Article I Section 4.2.2 of the treaty explicitly excludes from its definition of dumping the "placement of matter for a purpose other than

---

*This treaty came into effect in 1978, and as of January 2020, only 78 nations have ratified it. Alphabetically, the states to watch out for are Afghanistan, Algeria, Antigua and Barbuda, Argentina, Armenia, Australia, Austria, Bangladesh, Belarus, Belgium, Benin, Brazil, Bulgaria, Cabo Verde, Cameroon, Canada, Chile, China, Costa Rica, Cuba, Cyprus, Czech Republic, Democratic People's Republic of Korea, Denmark, Dominica, Egypt, Estonia, Finland, Germany, Ghana, Greece, Guatemala, Honduras, Hungary, India, Ireland, Italy, Japan, Kazakhstan, Kuwait, Kyrgyzstan, Lao People's Democratic Republic, Lithuania, Malawi, Mauritius, Mongolia, Netherlands, New Zealand, Nicaragua, Niger, Norway, Pakistan, Panama, Papua New Guinea, Poland, Republic of Korea, Romania, Russian Federation, Sao Tome and Principe, Slovakia, Slovenia, Solomon Islands, Spain, Sri Lanka, St. Lucia, St. Vincent and the Grenadines, State of Palestine, Sweden, Switzerland, Tajikistan, Tunisia, Ukraine, United Kingdom, United States of America, Uruguay, Uzbekistan, Vietnam, and Yemen.

the mere disposal thereof": in other words, if you've got a higher purpose than mere dumping—and you do—this treaty doesn't apply.

Whatever legal threats are brought against you, including those from individuals claiming losses due to inclement weather, you will be safest from prosecution in a country with lax environmental standards and limited extradition treaties. And given how the discussion around climate change has become a political, rather than scientific, issue—especially in the United States—there is a chance you could claim immunity under the "political offense exemption" found in many extradition treaties.

If you are apprehended, you have the advantage that, unlike a war crime and a crime against humanity (which, remember, is not what you're doing: this is a crime *with* and *for* humanity), there is no international court for environmental crimes. At worst, such disputes between states and non-state actors (like corporations, communities, or wildly self-realized individuals such as yourself) could be tried in an international court of human rights—and given how many lives you could (if all goes perfectly) save with this scheme, it's not at all clear what the verdict would be. Like many crimes discussed in this book, you could always try claiming the *Air Bud* defense: just as in the 1997 feature film there was nothing in the State of Washington's rules that specifically said a dog *couldn't* play in their basketball state finals, there's also nothing in current international law that specifically says someone *can't* enshroud Earth in a beneficial unofficial artificial white haze. And should any international authority create such a law in the aftermath of your heist, you should be immune: laws created after you do something in order to retroactively make that thing illegal are forbidden under Article 11, Paragraph 2 of the United Nations' own Universal Declaration of Human Rights.

Check and mate, rest of the world.

EXECUTIVE SUMMARY

INITIAL INVESTMENT	EXPECTED RETURN	ESTIMATED TIME UNTIL MATURITY
$7 billion up front, plus **$2 billion/year** ongoing	As this could affect the GDP of literally every nation on the planet, that sets the ceiling for the possible return of this plan at $87.8 trillion USD. Preventing even a 1% loss in productivity here gives a return of **$878 billion** per year.	**<10 years**

SOLVING ALL YOUR PROBLEMS BY DRILLING TO THE CENTER OF THE PLANET TO HOLD THE EARTH'S CORE HOSTAGE

I hate almost all rich people, but I think I'd be darling at it.
—Dorothy Parker (1958)

Hostage situations are a standard criminal trope: take someone prisoner, keep them fed and watered and comfortable, shyly ask them if they're familiar with Stockholm syndrome, negotiate a favorable trade for their release, and repeat ad nauseam until you've run out of either hostages or demands. This has been done before at human scale, and it is boring and it is tedious and it is *pedestrian*.

But it's never been done at the planetary scale before, which should at least keep things interesting.

BACKGROUND

Even though you've absolutely spent 100% of your life either on or near the Earth, it's possible there are still some things you don't know about it, so here's a quick primer. Like all great primers, ours begins by showing you what its subject would look like if you cut it in half:[*]

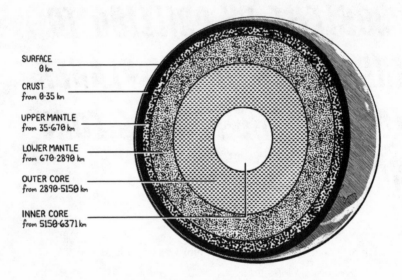

SURFACE
0 km

CRUST
from 0-35 km

UPPER MANTLE
from 35-670 km

LOWER MANTLE
from 670-2890 km

OUTER CORE
from 2890-5150 km

INNER CORE
from 5150-6371 km

The Earth, though not to scale, nor at actual size. To observe the Earth both to scale and at actual size, simply look up from this book.

We live on the surface of our planet: a very thin layer of solid rock—less than 1% of the Earth's volume—upon which rest oceans, buildings, and every living human who's not currently airborne or in space. When we dis-

---

*Cutting the Earth in half to verify this illustration is left as an exercise for the reader.

covered this noble layer—the home of *all known life in the entire universe*—we decided to call it "the crust."

The crust is a layer of largely solid rock that goes down 35km or so. It's not uniform, however. It's thickest—50km or more—at convergent tectonic boundaries, where one continental plate subducts beneath another: this is how you make mountains. And it's at its thinnest—10km or less—at oceanic divergent plate boundaries, where two tectonic plates are pulling apart. This allows the mantle underneath to reach the surface, which is how you make volcanoes.

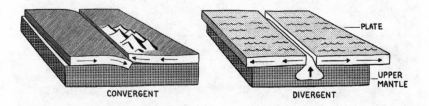

Convergent and divergent tectonic plates. Again, not to scale. To see them at scale, refer to your nearest mountain range or volcano.

The farther down you get into the Earth, the warmer it gets, for two reasons. The first is that the Earth was much, *much* warmer when it formed (frequent bombardment by meteorites and even other protoplanets as it formed gave it a ton of energy, so much so that its surface was mostly molten), and we've still got a lot of heat left over from that time. The second reason is that radioactive elements inside the Earth—things like uranium, thorium, and potassium—are constantly decaying, and they give off heat as part of that radioactive process. Of the around 47 trillion or so watts of energy that continuously flow from Earth into space, about half come from Earth's original primordial heat, with the other half coming from radioactive decay.

What's a Watt?

One watt is equal to one joule per second, and one joule is a relatively small measure of energy: it takes about 4.2 joules to raise 1 milliliter of water 1°C in temperature. Because of the 47 trillion joules of energy the Earth loses every second, the planet is cooling, but it's still a slow process: it'll take about 91 billion years for the Earth to get chilly enough to solidify completely. Don't worry about that having a negative impact on life here though: the sun will become large enough to sterilize and possibly consume the planet long before that ever happens! Chapter 9 has some spoilers on how that's going to go down.

This means that by the time you dig through the crust, you can expect to encounter temperatures getting as high as 400°C. (That's about 750°F.) As an intuitive comparison, the maximum temperature on most consumer ovens is 260°C, and 400°C is approaching "I can cook a cheesy flatbread in 60 seconds so that I might legally call it a Neapolitan pizza pie" territory.*

Beneath the crust is the upper mantle, a mostly solid bunch of silicate rocks that nevertheless behaves more like a thick fluid over geologic spans of time. I say "mostly solid" because the upper mantle is liquefied and mol-

*The Associazione Verace Pizza Napoletana, in their document describing the standards necessary to meet in order to get your pizza "Verace Pizza Napoletana" certified, calls for water that's between 16–22°C, for the dough to rise in a room at 23°C (plus or minus 10%), and for a pizza oven heated to 485°C. Now you know!

ten in some places, like under volcanoes, where it pushes up through the crust and reaches the surface. At the bottom of the upper mantle, some 670km underground, temperatures reach as high as 900°C—which for comparison's sake is the temperature of modern cremators, hot enough to reduce the average human body to just 2.4kg of ashes—and pressures here can be as high as 237,000 atmospheres.

This is a challenging environment to live in, let alone dig through, but let's finish learning about the Earth before we start to worry about punching a hole through it.

Beneath the upper mantle is the lower mantle. Temperatures here can get as high as 4,000°C—high enough to melt every element we know of, high enough to melt the rocks that the lower mantle is made out of, and easily high enough to vaporize diamonds. (I mention this because diamonds are typically used for drill bits, but let's deal with one problem at a time here.) Despite this heat, the rocks here are still mostly solid. That's thanks to the almost-inconceivable pressure of all the rock above—as high as 1.3 million times the atmospheric pressure you normally experience on Earth.

Dig through *that* and you'll reach the outer core: a liquid layer of metals, mostly iron and nickel, where things get the hottest, as warm as 7,700°C. That's above the boiling point of every known element, and the only reason the outer core isn't *itself* boiling is the pressure here, which reaches as high as 3.3 million atmospheres. The convection currents of the outer core produce the Earth's magnetic field, which, besides making compasses work, also protects us from most of the solar wind and thus makes life on Earth possible. If the magnetic field were, say, *suddenly stopped*, holes would form in the ozone layer as it began to be carried off into space by the solar wind. As these holes grew, lethal amounts of solar radiation would begin to bombard any life on the surface below. Over the long term (millions of years long, but still), the solar wind could strip away Earth's atmosphere almost entirely, which would cause air pressure to drop and the oceans to evaporate into space. This kills the planet, but it does tell us that keeping the magnetic field going is something everyone on Earth would pay good money to ensure.

Looks like you just found the perfect hostage.

Finally, beneath the liquid outer core is the inner core, a sphere of nickel-iron alloy about 70% the diameter of the moon. The inner core is similar in composition to the outer core, only it's solid, under even more pressure (3.6 million atmospheres), and cooler (5,400°C, though it can get up to 6,000°C at the boundary between the inner and outer cores). Again, while these temperatures are still easily hot enough to melt nickel, iron, or literally anything we know about in the universe, the extreme pressure keeps the inner core from liquefying. As the Earth cools, tiny particles of the liquid outer core solidify and fall like snow down to join the inner core, which causes it to grow slowly over time.

Compared to the constantly moving and magnetic-field-generating outer core, the solid inner core may seem a little dull, but don't dismiss it just yet. The reason these cores are mostly iron and nickel is that when the Earth was molten, the vast majority of the heavier elements on the planet sunk to the center in an event called "the iron catastrophe" (see sidebar on page 146), and there was a lot of iron and nickel to go around back them. But gold is a heavy metal too, and by looking at the composition of meteorites left over from the formation of our solar system (relatively gold-heavy) and comparing it to the composition of the crust (relatively gold-light), we can deduce that the gold that *should* be here has to have gone *somewhere*. Two possibilities suggest themselves: some of it could've vaporized and been carried off into space (possible at those primordial temperatures) while the rest of it must've sunk to the Earth's core. And this leads us to one inescapable conclusion: *there's gold in them thar inner and outer cores.* Up to 1.6 quadrillion tons of it, enough to coat all land on Earth in a layer of gold half a meter deep. All the gold ever mined in human history adds up to just 190,576 tons, and at the $58 per gram gold is worth in early 2021, that works out to about $11 trillion. This means the value of the core is sitting at over $80 sextillion USD in gold alone: that's a dollar sign, then an 8, and then 22 zeroes of pure profit.[*]

Another successful heist.

---

*This analysis assumes that unearthing 1.6 quadrillion tons of gold wouldn't crash the market, which it *definitely* would if you removed it all at once. So take your time! There's no (gold) rush!

The Iron Catastrophe

The time in Earth's history when most of its iron sunk to the core is called "the iron catastrophe," which is an *amazing* name—until you realize it uses "catastrophe" in the academic "sudden change" sense, and not in the more common "ruinous and deadly event" sense. As we've seen, without the iron catastrophe there'd be no magnetic field here on Earth, which means there'd likely be little atmosphere and no surface water . . . which means there'd be no life.

This makes the iron catastrophe quite possibly *responsible* for all known life in the universe, which is about as far from a ruinous and deadly event as one can go—unless, of course, you're going to be one of those supervillains with a "humans are the real virus" ideology, in which case: congratulations on finding your new favorite "fun villainous monologue" factoid!

And yes, even at its maximum estimation, all that gold only adds up to about one part per million of the core itself, but *still*. Almost all of Earth's gold—99.5% of it—is missing, some of it's inside the inner and outer cores, and if you can reach it, all that gold is yours (as is all the silver, platinum, and other precious metals that are trapped there too). You've just found the perfect hostage, the perfect motive, *and* the perfect fallback funding scheme.

At this point we should note that there are some other sub-layers we've not discussed, particularly at boundaries between layers. But what's here is the high-level shape of the planet we all share, and which you now intend to hold hostage and relieve of its secret gold.

Now all you need to do is dig through it.

THE INFERIOR PLANS OF LESSER MINDS

The world's deepest hole, as of this writing, is the now-abandoned Kola Superdeep Borehole, located on the Kola Peninsula in Russia, north of the Arctic Circle. It's a hole 23 centimeters (cm) in diameter, and it was started in May 1970 with a target depth of 15,000m. By 1989, Soviet scientists had reached a depth of 12,262m, but they found they were unable to make further progress due to a few related issues. The first was that temperatures were increasing faster than they'd expected. They'd expected to encounter temperatures of around 100°C at that depth but encountered 180°C heat instead, which was damaging their equipment. That, combined with the type of rock found and the pressure at those depths, was causing the rock to behave in a way that was almost *plastic*. Whenever the drill bit was removed for maintenance or repair, rocks would move into the hole to fill it. Attempts to dig deeper were made for years, but no hole ever made it farther than 12,262m, and the scientists were forced to conclude that there was simply no technology available at the time that could push any deeper. The Soviet Union dissolved in 1991 in an unrelated event, drilling stopped in 1992, the site was shut down, and the surface-level opening to the hole was welded closed in 1995. Today, the drill site is an abandoned and crumbling ruin, and that still-world-record-holding maximum depth, 12,262m, is less than *0.2%* of the way to the Earth's center, some 6,371km below.

So, that's a concern.

But that was back in the '90s, and we humans have continued to dig holes since! The International Ocean Discovery Program (IODP) has a plan to dig through the thinner oceanic crust, hoping to break through to the mantle and recover the first sample of it taken in place—but this project, estimated to cost $1 billion USD, has not yet been successful. Still, a ship built for the project, the *Chikyū, has* briefly held the world record for deepest oceanic hole (7,740m below sea level!), until it was surpassed by the *Deepwater Horizon* drilling rig, which dug a hole 10,683m below sea level and then exploded.[*]

The evidence here all points to one depressing conclusion: the deepest holes humanity has ever made don't go nearly far enough, and they've already reached the point where things get too hot—and too plastic—to continue.

But these holes were all dug not by supervillains chasing lost gold but by scientists, a group largely constrained by their "ethical principles" and "socially accepted morals." To a supervillain, the solution here is obvious. If the problem is that the rocks are so hot that they're damaging equipment and flowing into the hole, why not simply make a hole wide enough that some slight movement isn't catastrophic, and cool enough so the rocks are all hardened into place? Why not simply abandon the tiny, 23cm-diameter boreholes of the Soviets and the similarly sized drill holes of the IODP, and instead think of something bigger? Something bolder?

Something like a colossal open-pit mine?

Such a mine would minimize the effects of rocks shifting by giving them a lot more room to shift—and us a lot more time to react—before they become a problem. You could keep those rocks cool and rigid with one of the most convenient coolants we have: cold liquid water. On contact with hot rocks or magma, water turns to steam, carrying that heat up and away into the atmosphere, where it can disperse naturally—while at the same time cooling the rocks so that they remain both solid enough to drill and

---

[*]Okay, *technically* the explosion wasn't quite so immediate. *Deepwater Horizon* finished drilling the world-record-setting hole in September 2009 and then experienced a blowout in April 2010 . . . before sinking two days later while setting another world record, this time for the largest marine oil spill in human history.

rigid enough to stay in place. It would take an incredible amount of water, but lucky for us, Earth's surface is 71% covered with the stuff!

So if you build a sufficiently large open-pit mine next to the ocean and use a dam to allow water to flow into the pit to cool the rocks as needed, then you'll be the proud owner of a mine that allows you to reach greater depths, both literal *and* metaphorical, than anyone else in history! This scheme has the added benefit that, if we're clever, we can use the steam that's generated by cooling all that hot rock and magma to spin turbines, which could then generate more power for drilling. You'll build a steam engine that's powered by *the primordial and nigh-inexhaustible heat of the Earth herself.*

The combined might of science and villainy teaches us that the problems encountered with an incredibly skinny hole can be mitigated by an incredibly wide one.

The exact dimensions of open-pit mines vary depending on what's being mined, but they're all shaped like irregular cones, with the biggest part at ground level and the smallest part at the bottom of the pit. The open-pit mine that's both the world's largest and deepest is the Bingham

Canyon copper mine in Utah: it's been in use since 1906, and in that time it has produced a hole in the Earth's crust that's 4km wide and 1.2km deep.* Using those dimensions as a rough guide produces the following chart:

AN OPEN PIT DEEP ENOUGH TO REACH THE BOTTOM OF EARTH'S...	...WOULD HAVE TO BE AT LEAST THIS DEEP...	...AND AT LEAST THIS WIDE:
Crust	35km	116km
Upper mantle	670km	2,215km
Lower mantle	2,890km	9,554km
Outer core	5,150km	17,025km
Inner core	6,371km	21,061km

A sinister and threatening collection of numbers that, should someone ask you what you're reading, you should quickly cover up and say, "Nothing!"

. . . and here we have another problem. Just reaching the bottom of the crust needs a hole over five times the length of the island of Manhattan, dozens of times wider than any other hole made by humanity, and easily large enough to be seen from space. Reaching the bottom of the lower man-

---

*This produces a ratio of width to depth of 3.31: we get 1km deeper for every 3.31km wider our hole gets. Every open-pit mine has its own ratio of width to depth, but if we look at the deepest ones, we see they're all roughly similar. The second-deepest open-pit mine in the world is the Chuquicamata copper mine in Chile, with a width of 3,000m, a depth of 900m, and a ratio of 3.33. The third-deepest, the Escondida copper mine, also in Chile, is 2,700m wide and 645m deep, which gives a ratio of 4.19. The fourth-deepest, the Udachny diamond mine in Russia, has a width of 2,000m, a depth of 640m, and a ratio of 3.13. While it's true that these mines were built to "recover minerals" and not "dig as deep as possible so that one might hold the Earth's core hostage," these do form a reasonable rough estimate for how big our hole will have to be.

tle would require a hole so huge that its opening would encompass 75% of the Earth's diameter, and to do the same with the outer and inner cores requires holes that are wider than the Earth itself.

Even if you could turn almost half the Earth into an open-pit mine cooled by seawater, the steam created by cooling a pit that size would effectively boil the oceans and turn the Earth into a sauna, destroying the climate, collapsing food chains, *and* threatening all life on the planet—and that's *before* you even reach the hostage-taking phase, let alone the part where you plunder forbidden gold! Things get even bleaker once you take into account the responses from the governments you'd upset by turning their countries into hole; the almost inconceivable amount of time, energy, and money required to move that much matter; where you'd put all that rock once you dug it up; or the true, objective inability for anyone, no matter how well funded, ambitious, or self-realized, to possibly dig a hole this huge.

So.

That's another concern.

YOUR PLAN

It pains me to say this, but . . . there is absolutely no way, given current technology, for anyone to dig a hole to the center of the Earth no matter how well funded they are, *even if they drain the world's oceans in the attempt.* We have reached the point where your ambition has outpaced even *my* wildest plans, most villainous schemes, and more importantly strongest and most heat-resistant materials. Heck, we're actually closer to *immortal humans* (see Chapter 8) than we are to tunneling to the Earth's core. It's unachievable. Impossible. There's simply no way forward.

THE DOWNSIDES

It's truly, truly hopeless. It's hard for me to admit it, but even the maddest science can't realize every ambition.

I'm sorry. There's nothing more I can do.

... for this plan, anyway!

But every good villain always has a Plan B, one that snatches victory from the jaws of defeat. And heck, if you've got your heart set on digging a hole, making some demands, and becoming richer than Midas and Gates and Luthor in the process—*who am I to stop you?*

YOUR PLAN B

You're going to sidestep the issues of heat and pressure in the Earth's core by staying safely inside the crust, within the depth range of holes we already know how to dig. And you're going to sidestep the issues of legality that tend to surround schemes to take the Earth's core hostage by instead *legally* selling access to your hole to large corporations and the megarich, who will happily pay through their noses for the privilege. Why?

Because instead of digging down, you're going to dig *sideways*. Instead of mining gold, you're going to mine *information*. And unlike even the lost gold of the Earth's core, this mine is practically *inexhaustible*.

It all has to do with stock trading. In the mid-twentieth century, stock exchanges had trading floors, which were actual, physical floors where offers to buy and sell were shouted, out loud, to other traders. It was noisy and chaotic, but it ensured everyone on the trading floor had, in theory, equal access to the same information. Those floor traders were later supplemented by telephone trading, and then almost entirely replaced by electronic trading, which is how most stock exchanges operate today. At the time, both telephone and electronic trading could be pitched as simply a higher-tech version of the same floor trading that already existed, but they also did something more subtle: they moved trading from the trading floor to outside the exchanges themselves, where everyone might not have access to the same information.

Turns out, there's money to be made from that.

Obviously, if you know information about the stock market that nobody else does, you can use that to make money. Before telecommunications, this was accomplished by having someone *inside* a publicly traded company leak tips: say, that an annual report was about to come out showing unexpected losses. You could sell your stock before anyone else suspected there was even a reason to sell, and thereby avoid the losses everyone else would take—or, going the other way, buy a stock before the rest of the world knew it was about to explode. This is called "insider trading," and it's illegal.[*]

But the thing is, it's generally *not* illegal to profit from information that's public in one part of the world but still unknown someplace else.

For example, if you were in San Francisco in September of 1860 and got some world-changing news from New York—say, delivered by the Pony Express and their relay of horses, which before the invention of the telegraph was the fastest way to send information across America—then it was *entirely legal* for you to profit from that information, even if nobody else around you knew about it. This window of time might have been short—lasting only a few hours, until that next pony rode into town telling someone else the same thing—but it was yours to exploit if you could.

Today, of course, the constant stream of information flying around the world has ensured that the hours-long windows of information privilege is gone. But small windows wherein you know something before anyone nearby do still exist, and what's nice about this scheme is that it still works even if your time advantage is measured in only *milliseconds*. If you can get stock information a fraction of a second before anyone else, *and* you can buy or sell before that window closes and the information becomes available to everyone else, you can make mad bank.

That's obviously far too short a time for any human to react, so we've built programs to trade for us, making decisions automatically based on certain criteria. The ultimate conclusion of this process is what's called

---

*Obviously as supervillains we all know that being in trouble is a fake idea, but it's always nice to avoid *unnecessary* trouble with the law.

"high-frequency trading," wherein programs buy a stock, hold it for possibly only a few milliseconds, and then sell it again: it's that old adage of buying low and selling high, now realized dozens or possibly hundreds of times every second, achieved by software moving as fast as possible to exploit information that competing software doesn't yet know. As a simple example, if your program is the only one in Toronto that knows there's someone in New York looking to buy a certain stock at, say, $10, then it could purchase shares of that stock in Toronto, where it's currently trading for $9.95, and make an all-but-guaranteed five-cent profit per share when that buy order from New York arrives a few milliseconds later. It's not a ton of money, but do it several times per second, millions of times per day, and you can produce billions in profit with almost no risk—because effectively, your program is *seeing into the future*, operating off knowledge that won't arrive for everyone else until after you've acted upon it.

All you need now is a way to know about what's happening in stock markets around the world before anyone else does.

For a while, information between stock exchanges was transmitted via fiber-optic cable, where information is transmitted as light itself—which meant that a more direct fiber line from a stock market to a trader could be a gold mine. In 2010, a company named Spread Networks approached Wall Street financial markets, revealing that they'd spent the past year building a new fiber line—*in secret*—between the Chicago Mercantile Exchange and the data center beside the Nasdaq's New Jersey stock exchange. Unlike existing telecommunication lines that took the path of least expense—doing reasonable things like following railroad lines and going around mountains—this new 38-millimeter (mm) plastic tube of underground fiber-optic cable went in the straightest line possible, even passing *through* mountains instead of going around them. In order to keep secrecy, a variety of shell corporations with dull names like "Northeastern ITS" and "Job 8" were used for construction, and workers contracted to these companies weren't told *why* they were building this tunnel. They were simply told that "the client" wanted it as straight as possible and that they were to report to their supervisors anyone who hung around the site or asked too many questions about it. The secrecy worked: the 1,331km-long tunnel was built without anyone realizing what it was for, and when it was completed, this new fiber-optic line was over 160km shorter than any other connection between the two cities. Spread Networks could now move information between New Jersey and Chicago faster than anyone else in the universe.

"Round-trip travel time from Chicago to New Jersey has been cut to 13 milliseconds," was the pitch: over 1.5 milliseconds faster than the next-fastest fiber route. The cable had enough bandwidth for 200 people to have their own simultaneous connection, and they found that companies were not only willing to pay for it—they were willing to pay more to ensure that others *didn't* have it. The cost for access started at $14 million USD* for a five-year lease, raising to almost $20 million if the client didn't pay it all up front, which meant at least $2.8 billion of revenue in the first five years alone.

Building the tunnel itself cost just $300 million.

This is worth highlighting: this scheme has already been pulled off before, and it was both legal† and profitable.

But while fiber-optic cables transmit information *with* light, it doesn't travel as fast *as* light. Light travels at different speeds through different media, and fiber-optic cables slow it to around 70% of its max speed in a vacuum: still wildly fast, but not fast enough that you can't beat it with something better. That's why today information between stock exchanges is often transmitted by microwave emitters, which have the limitation that they only work along a line of sight—each transmitter has to be precisely aimed at the next receiver along a chain and have a consistently

---

*It cost $10.6 million for access to the line, but the clients would need to buy and maintain their own signal amplifiers, housed at ground level, at 13 sites along the fiber route, which brought total costs to about $14 million a head.

†It was legal due to the 400 or so deals made with the small townships, counties, and private entities to allow Spread Networks to tunnel across their land. If you're a villain who doesn't care about laws, dig in secret!

clear view of it, meaning inclement weather can disable them—but also the advantage that they can transmit information at more than 99% the speed of light, which is another way of saying they send information at over 99% of the maximum speed it's possible for *anything* to travel in this universe.

Unsurprisingly, these microwave transmission chains have *also* been built in secret in order to make mad bank. A microwave link between London, England, and Frankfurt, Germany, was built in secret by a company called Perseus Telecom—and the only reason this is known publicly is that in 2013 a rival company built their own microwave link and publicly solicited customers, which caused Perseus to reveal their hitherto-exclusively-privately-used network and do the same.

But even these state-of-the-art, as-fast-as-humanly-possible transmission networks of underground fiber-optic cable and above-ground microwave transmitters share the same weakness: *they follow the curvature of the Earth*. Remember: the ground beneath you may look flat, but it has a very slight curve because we're all standing on a colossal giant ball of rock and metal tearing through the Milky Way. The Earth's curve is not usually a factor at human scale, but when you get to more global scales—say, the distance between two cities—it becomes a very important one. And while the shortest distance between two points is always a straight line, a line that follows the surface of the Earth isn't straight: it's a curve. The truly straight line between two cities—and the stock exchanges therein—cuts through the Earth on its way: say, via a bespoke tunnel that finally allows information to travel between these two locations as fast as it possibly can.

You are limited by distance here: try to connect areas too far apart, and you'll go deeper than the 12km or so limit that we humans have been able to dig. That dictates that any two places on Earth more than 790km or so apart are going to hit that limit on our digging technology, and you may run into difficulties before then. But it just so happens that the ninth-largest stock exchange in the world—with a market capitalization of $3.1 trillion USD in 2020—is the Toronto Stock Exchange in Toronto, Canada.

And the world's largest stock exchange—with a 2020 market capitalization of a whopping $25 trillion—is the New York Stock Exchange, which is just 550km away by land—or only *549.8km* away through direct subterranean tunnel. And such a hole, at its lowest, would be just 5.9km below ground: less than half the depth of the deepest hole, and well within the range of even Soviet-era drilling.

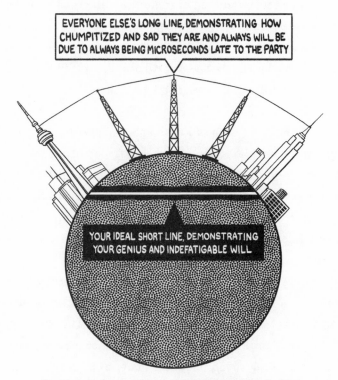

The triumph of your villainous scheme. Again: both the Earth,
and your brilliance, are not shown to scale.

Run fiber through such a hole and your round-trip time between Toronto and New York is just 5.3 milliseconds. Your closest competitor is likely Crosslake Fiber, who in 2019 ran fiber-optic cable across the bottom of Lake Ontario to minimize Toronto/NY transmission times, and the best round-trip speed *they* got was 9 milliseconds. Heck, your fiber will be 0.002 milliseconds faster than the round-trip transmission time of even an ideal straight-line surface-level fiber connection, which, and this only helps you further, *doesn't actually exist*. Put microwave emitters into that hole and this time drops down to just 3.7 milliseconds—and since your tunnel is protected from the weather, it'll still work in heavy rain and snow, which interfere with any microwave transmitters on the surface. (And if you wanted to beat Spread Networks at their *own* game, a direct tunnel through the Earth between their data center in Aurora, Illinois, and the New York Stock Exchange would be challenging enough to take you to the limits of what's possible—it'd be 1,200km long and 12.18km underground at its deepest, 99% as deep as the Soviets got—but it'd beat the Spread Networks fiber-optic line by 1.4 milliseconds (ms), and by another 3.5ms on top of that if you use microwaves.)

You, and you alone, will now know what's happening at the opposite exchange—and the trillions of dollars being traded inside them—a fraction of a second before anyone else on Earth. And while gold mines run out and even the Earth's core could be mined to exhaustion, trade is as old as humanity. The very laws of physics themselves are preventing anyone from transmitting information between these two cities as fast as you—unless they too manage to build a similarly venturesome hole.*

The two longest tunnels on Earth are the Delaware Aqueduct in America (137km long, 4.1m in diameter) and the Päijänne Water Tunnel in Finland (120km long, 4.5m in diameter, dug through solid bedrock). Your tunnel is longer than those, but given the width of these tunnels, you'll actually be excavating *significantly* less material than they did. With a 23cm diameter tunnel between the Toronto and NY exchanges, that's just 22,843m³ of rock and soil that would need to be removed, compared to over 1.8 million cubic meters required to be moved for each of the two water tunnels. So while this scheme will take time and money, it's not impossible.

---

*Better buy up all the copies of this book, just to be safe.

In fact, it's better than just "not impossible." It's *bold*. It's *audacious*. And it's legal, or at least, legal enough that the money you make for yourself and other powerful people should be able to push it over the finish line. Furthermore, since it involves digging a secret tunnel under the Earth's surface—and across an international border—solely to make yourself rich by indefinitely manipulating the stock market to your own advantage by exploiting information nobody else around you has, it's definitely super-villainous.

See? This book has already earned you back its cover price, and you're not even halfway through!

POSSIBLE REPERCUSSIONS IF YOU'RE CAUGHT

Thus far, while this style of high-frequency trading has been criticized in American and many other international stock markets, it has not been made illegal, so *welcome to this viable business scheme.* There are no repercussions for being caught doing anything, as long as everything you're doing is legal!

But even if it's *legal*, that doesn't mean it'll always be *possible*. There is one stock exchange—the Investors Exchange, or IEX—that runs in New Jersey, and that was specifically designed from the ground up to mitigate any speed advantages that you, and any high-frequency traders, could exploit. As such, the IEX has its own direct fiber connections to other exchanges in New York, which take *at most* 320 microseconds to transmit information to the world, while the fiber-optic cable that every trader has to use is longer than it needs to be: 61km longer, in fact. Those extra kilometers of cable, tightly coiled and packed into a box inside the IEX server room, ensures an unavoidable physics-imposed delay of *at least* 350 microseconds. By ensuring it's impossible for any trader to beat the speed of the market itself, the IEX has produced a way to effectively—and automatically—foil our "get information faster than anyone else" stock-market scheme.*

Luckily for you, since launching in 2013, the IEX has never captured more than 3.4% of the market share, and recent trading has it below 2%.

Apparently, American stock traders aren't especially interested in killing the golden goose that is making them—and soon, you—so much money. Despite the unfairness of this system—and the fact that some high-frequency traders have bragged about going half a decade without losing money on even a single day of trading (something effectively impossible unless you know a lot of things other people don't)—this remains a way of making money that we've all decided, as a society, is fine. It's fine! We should all stop worrying about it! One *could* argue that this sort of trading functions as a drag on trade, increasing prices and parasitically siphoning as much as $22 billion USD a year in profit for itself without doing any productive work for civilization while, simultaneously, incentivizing the construction of absurd and pointless telecommunication megaprojects, but the people arguing *that* tend not to have their own private secret tunnels through the Earth's crust, so what do they know?

Speaking of tunnels, these can of course be built legally, even across international borders, and the fact that your tunnel—nay, data communication corridor—is far too small to permit human trafficking is an advantage here. And if you do decide to just start digging without the proper permits? Well, laying cable without permission is hardly life-in-prison territory. United States Code, Title 18, Part 1, Chapter 27, Section 555—"Border tunnels and passages"—says, "Any person who knowingly constructs or finances the construction of a tunnel or subterranean passage that crosses the international border between the United States and another country, other than a lawfully authorized tunnel or passage known to the Secretary

---

*It's not a perfect solution, of course: the reality of the world is that some people will always get information before others do. But this does limit the number of trades that anyone can make while this information is still propagating, which limits the profit that can be made in this time, thereby disincentivizing the scheme.

of Homeland Security and subject to inspection by Immigration and Customs Enforcement, shall be fined under this title and imprisoned for not more than 20 years." That's 20 years worst case, and that's under a law that's more focused on people who smuggle "an alien [sadly the word is used here in its much more disappointing 'anyone who's not American' sense], goods, controlled substances, weapons of mass destruction, or a member of a terrorist organization" than it is someone transmitting such staid information as publicly available stock prices. Besides, tunnels are very difficult to detect—even those that are big enough to send people through—so once built it's relatively safe.

And finally, there is some precedent for secret holes being forgiven, at least when they enter and exit within a single country.

In early 2015, a tunnel 3m deep, 1m wide, 2m tall, and more than 10m long was found in Toronto, near the campus of York University. It was only discovered when its secret entrance, hidden under a sheet of corrugated steel itself hidden under a pile of dirt, was uncovered by a park worker. It was expertly constructed: the ceiling was supported by thick wooden beams, it had a sump pump to keep water out, moisture-proof lights, and even a generator in its own separate and soundproofed underground chamber some 10m away, which powered the tunnel via buried extension cords. The allure of this mysterious secret tunnel—which clearly took effort to execute but seemed pointless in intent—ensured that it was reported on in Canada, America, and even the UK.

Later, after the police had filled in the hole and assured the public that no crime had been committed, Elton McDonald, a talented 22-year-old construction worker, revealed himself as the one responsible for the tunnel—with the photographs of its construction to prove it. He said the tunnel was built as a chill-out hole by and for himself and his buddy: someplace they could hang out that was cooler in the summer and warmer in the winter. It was a place they could watch movies on a laptop or listen to music, and while they'd had barbecues inside the hole, it wasn't finished yet: there were two larger rooms planned, and a television to be installed in one of them down the road. McDonald had started work on the hole in 2013 as a hobby, and work on the project had only ever been part-time before it was discovered.

He expressed disappointment that the police had filled in his party hole but vowed to "do it again": this time, on his own property—once he could afford some.

Hole engineer Elton McDonald will not be stopped, and neither will you.

EXECUTIVE SUMMARY

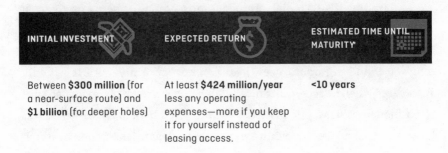

INITIAL INVESTMENT	EXPECTED RETURN	ESTIMATED TIME UNTIL MATURITY*
Between **$300 million** (for a near-surface route) and **$1 billion** (for deeper holes)	At least **$424 million/year** less any operating expenses—more if you keep it for yourself instead of leasing access.	**<10 years**

TIME TRAVEL

> If you must know anything, know that the hardest task is to live
> only once.
>
> —Ocean Vuong (2016)

Despite my best efforts and all my previous work, I have not yet been able to successfully induce even a single instance of time travel in any temporal direction other than forward, and only then at a rate of one second per second, relative to local time.

If that changes, I will go back in time and adjust this manuscript before it goes to press, thereby ensuring that every copy of this book is automatically updated.

Please check this chapter periodically.

EXECUTIVE SUMMARY

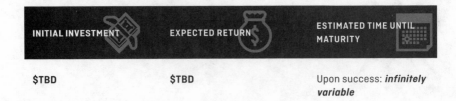

INITIAL INVESTMENT	EXPECTED RETURN	ESTIMATED TIME UNTIL MATURITY
$TBD	$TBD	Upon success: *infinitely variable*

7

DESTROYING THE INTERNET TO SAVE US ALL

> We have too many cell phones. We've got too many Internets. We
> have got to get rid of those machines.
>
> —*Ray Bradbury (2010)*

In the space of a generation, the internet has changed—even for us super-villains!—from something you eagerly anticipated getting to spend some time with to something that for many of us is an always-on, omnipresent, and absolutely mandatory part of life. We interact with internet-connected machines at work and at home, and when we're outside we carry a portable internet-connected machine in our pockets *just in case*. We use the internet to make money and to spend it, to make new friends and to dunk on strangers, to get food delivered and to get paid to deliver food. It's where we keep up with our friends and family, where we find companionship and love and sex, and it's where we post our jokes about hating the bad screen at work and looking forward to staring at the good screen at home. We log on to social media to share our unique and beautiful thoughts with the world, and it's in that exact same space where we realize, too late, that the price of doing so is to be regularly exposed to the unsolicited thoughts of the suckiest fools on Earth.

The internet is a place of gossip and misinformation, hate mobs and brigades, anonymous threats and crowdsourced harassment, where people who despise you will go digging through decades of old posts just for the *chance* they'll find something to discredit you with. It's a place of confidence tricks and scams, of chain letters and multilevel marketing, of manipulation and innuendo and larceny and surveillance. It's a place where everything is saved and nothing is forgotten, but which still managed to turn your cousins into conspiracy theorists and your nephews into Nazis.

And while meeker people might just sit back and let those things happen because "the internet is important" and "we can't live without it now" and "what am I supposed to do about it," you're not meeker people. You're a supervillain, damn it, and no computer network made by (and originally for) nerds is going to push *you* around. Surely at some point you've wondered whether things wouldn't be better if one day, somehow, this whole "internet" thing just . . . *went away.*

Impossible, right?

Let's find out.

BACKGROUND

We think of the internet as being somehow ethereal, above the petty reality the rest of us live in. But even the fanciest website, streaming platform, subscription service, or villains-only dating app is just software running on physical hardware. No matter how much we conceptually abstract away from the actual machines—and we have, with many internet services today running on virtual servers that are themselves running in "the cloud," a collection of computers acting as a globally distributed computing platform available to anyone who'll pay—every single thing you've ever read, watched, or listened to online came from particular pieces of physical hardware responding to your individual request. The basics of the internet have not changed since its invention: it used to be a short, single connection running between two institutions, and now it's over 1,200,000km of cable connecting billions of computers across every continent on the planet, but it's still just signals being transmitted to and from machines, and those machines are just electrified metal we built in order to add up numbers faster than we could in our heads.

Here's how the internet works.

Early computers used to talk to each other directly: your machine would connect directly to another machine, like having a conversation one-on-one with someone in a locked room. This was obviously unsatisfactory, so then we came up with larger networks in which we could easily talk to more than one machine at the same time. Now, instead of being in a locked room with one machine, it was more analogous to all our computers hanging out in a big swimming pool, where they could see and chat with whomever they wanted. This is an excellent analogy that I am not accepting feedback on at this time.

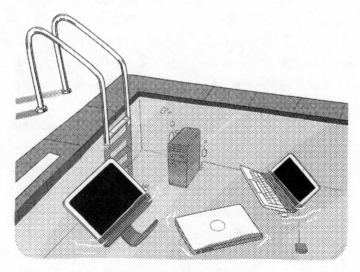

One excellent analogy.

This wasn't yet the internet, however, because these early networks solved the problem of "how does person A talk to person B when they live in different cities" in a more primitive way than the internet does. Earlier computer networks, inspired by the telephone networks that preceded them, used a system in which there were *set routes* for data. For example, if you were in Los Angeles back in the day and wanted to check in with my headquarters in New York, your phone call would be routed by an LA operator (initially, a human being who would physically connect your call by moving wires in a switchboard) to, for example, a similar hub in Las Vegas, which then patched you through to the Albuquerque hub, which connected you to Kansas City, then to Chicago, and then to New York, where you'd finally be connected to the line running directly to the phone on my desk. At every step along the way, each operator knew that a call from *that* city to *this* city would take *this* set route through the network, such as the LA to New York route we saw. And when everything works perfectly, such a system works great—but if Albuquerque goes down, then *all* connections running through that city are severed until the routes are rebuilt, manually, by a human being. This means that the more popular a hub is, the larger a liability it becomes, as a failure there can now break a large part of the system. Any computer network built like this is *brittle*, and the U.S. government wanted a stronger communication network than that. They wanted a network that would still work even if Albuquerque were nuked. Heck, they wanted a communication network that would still work even if *large parts of the United States* were nuked.

This is what led to ARPANET, the precursor to the internet, where redundancy is baked into its fundamental structure. Rather than an overseer determining fixed routes, messages are sent through a process called "packet switching." With packet switching, instead of a whole message taking a preset route, any communication is broken up into tiny parts—packets—each of which can take their own unique path to their destination. This system doesn't guarantee that every packet will arrive—much less that they will actually arrive in order—but the internet protocols solve this by wrapping each packet in a virtual envelope to identify them (effectively saying, "Hey, what's up, I'm part 7 of 9 of *this* particular message from *that* particular computer") and then ensuring the recipient can request that any lost packets be resent. This way, if Albuquerque is nuked (or, *usually* more likely, if a computer you were using to transmit your message through Albuquerque goes down), then you can still get your message via some alternate path. Any malfunctioning computer gets routed around automatically, allowing the system to stay connected even if large parts of it are destroyed, and even a fragment of a message arriving can set in motion a process that recovers the whole.

And that means we internet-destruction-minded supervillains are left trying to figure out a way to take down a resilient network of machines *literally designed to survive a nuclear bomb.* Sounds like our work is cut out for us . . .

. . . is what we'd be saying, if it weren't for the past several decades of civilian internet development!

Mathematical Proofs About Human Communication

Here's a fun fact about resending lost data: you can never be 100% certain that both people, communicating on a network where information *might* get lost, share the same information—and this holds even if nothing ever actually gets lost!

Here's an example. Let's say you want to send the very important message "hey baby do u like villainous ppl bc i am v villainous and SOMETIMES into kissing ON OCCASION!!!." You could send that message and, being clever, ask the recipient to send you back an acknowledgment so you'll know your message made it through. Easy, right? And let's say everything goes as planned: they get your message and send you back an acknowledgment. But when it arrives, while *you* now know that they got your villain-maybe-available-for-kissin' message, *they* don't know that you know that, because there's a chance their acknowledgment might not have made it through! So you send them a "thanks, I got your acknowledgment" message, which means *they* now know that *you* know that *they* got your initial message, but *you* don't know that *they* know that *you* know that it went through. And as this process continues, it becomes clear that the only two options you have—and this is proved through the ruthless mathematics of distributed computing—are to keep sending acknowledgments of acknowledgments (of acknowledgments of acknowledgments) for the rest of time, ever striving for the apparently simple but literally hopeless goal of one set of facts about the state of the world being perfectly understood by two minds . . . or to instead accept that we can never truly know each other in this fallen realm, and move on with our lives.

Similar distributed computing proofs have shown that there's no way to reach consensus (which is to say, have everyone in the network agree on something that happened) if more than a third of the actors in that network are willing to lie about it. Anyway, that's just another fun fact about computer networks that could maybe apply to human societies, who can say?

THE INFERIOR PLANS OF LESSER MINDS

Because of the internet's resilient routing, we know that cutting a few wires—or even melting a whole city's worth of wires by dropping a nuclear bomb on 'em—wouldn't be enough to destroy the internet, and the remaining machines would route around the damage. But while it *began* life as a decentralized network of independent machines, the past few decades have seen the internet become increasingly centralized in the hands of just a few corporations.

In the previous section I mentioned "the cloud": this is simply a bunch of computers networked together so that they present as a single computing platform. Using a cloud computing platform, you could access the power of one physical machine—like the one you have on your desk—or, for a few dollars more, you could harness the combined computing power of *hundreds* of these machines all working in parallel. For people in charge of keeping websites up, it's a very attractive proposition. With one physical computer, a corrupted hard drive means disaster and a broken website. But with your data backed up across a massively redundant array of machines in the cloud that all work together to present as a single virtual machine, a corrupted hard drive merely means that one computer is taken out of service, the others take up the slack, and you never even notice. For a monthly fee (which you can adjust at your convenience by renting more or less computing power!), you can have someone *else* be responsible for all that hard, down-in-the-muck-of-reality work maintaining and repairing physical hardware. And free as you are with this more resilient virtual platform, you can now focus on more interesting high-level stuff like "what should I post on my secretly villainous recipe blog today," secure in the knowledge that everything else will *just work*. Because of that attractive proposition, many of the world's most popular websites now run on one or more private cloud services, which, due to their complexity and startup costs, are generally offered by colossal companies like Amazon and Google.

And that makes them *targets*.

Take out Amazon's and Google's cloud platforms, and you not only damage those companies, you damage a large chunk of the existing and mainstream internet. (If you're reading this book while inside Google or Amazon, just *be cool*, act normal, maybe don't leave this page open for too long.)

You can get a taste of what that would be like semiregularly: when Amazon's cloud hosting service, AWS, suffered downtime in just one of its 23 regional data centers in November 2020, it affected internet apps as varied as graphic design software from Adobe, cryptocurrency trading on Coinbase, photo sharing on Flickr, streaming media on Roku, podcasts like *Radiolab*, home phone service over the internet through Vonage, robot vacuums through iRobot, password management through 1Password, the news itself through the websites of *The Philadelphia Inquirer* and *The Washington Post*—and that's not even considering the countless smaller sites and services that were knocked offline. Even Amazon's own website that reports on the health and status of the AWS network to subscribers was unreachable.

In other words, a single data center outage broke a large part of the internet. Even though it had been built to withstand the force of a *nuclear* *bomb*, it turns out that the forces of capitalism—of advertising, economies of scale, and near-monopolies—are in some ways even stronger than nukes. As a result, large parts of the internet are once again brittle, relying as they do on a few single services with their single points of failure. It's worth highlighting that this outage at AWS, like most cloud-hosting outages, was *unintentional*: this one was caused by a mere routine upgrade that went off the rails.* This points to a remarkable conclusion: some of the most motivated people online who are working for some of the richest companies in the world, and who, thanks to the runaway success of cloud computing platforms, have seen their jobs become in a very real way to *stop the internet from breaking*, can still fail at their jobs *by accident*.

If you'd like to go even further in simulating your success, you could follow in the footsteps of investigative journalist Kashmir Hill, who in July 2020 reported in *The New York Times* on her efforts to use an internet from which she'd blocked Google, Amazon, Facebook, Apple, and Microsoft.† (Besides not using their hardware, she'd blackholed their IP addresses, effectively making it appear—just to her computers—that every one of these companies had gone completely offline.) "Amazon and Google were the hardest companies to avoid by far," she wrote, noting how so much of the internet was hosted on their machines: Amazon Prime Video streaming was out, obviously, but so too was their competitor Netflix, which used AWS hosting. She found even seemingly unrelated sites also broke in unexpected ways: for example, blocking Google also blocked their commonly used captcha service, which meant she couldn't log in to any websites that used it because she could no longer prove that she was a human. It

---

*If you're interested in the specifics, here's what happened: Each front-end server in AWS's cloud maintains a connection to every other computer in the fleet, used to communicate and coordinate across servers. When the new servers were added, this meant that the number of connections required for each server to talk to every other server now exceeded the amount their operating systems could handle, causing a cascade failure across the entire fleet.

†Hill's *Times* article summarized the results of a delightful and dystopian series of articles she'd written for *Gizmodo*, describing the experience as she blocked each company individually, and then, at the end, all at once.

also broke any app that used these company's map platforms, including Uber and Lyft. Even using a privacy-focused alternative search engine to Google—DuckDuckGo—failed, because it was hosted by Amazon's AWS platform: she fell back to Ask.com, the descendant of OG butler-themed search engine Ask Jeeves. Hill eventually resorted to buying a physical map to get around, noting, "I came to think of Amazon and Google as the providers of the very infrastructure of the internet, so embedded in the architecture of the digital world that even their competitors had to rely on their services."

All this is to say: the modern internet is vulnerable in a way it wasn't before and was never intended to be. Your success here would fundamentally change both what the world looks like and how it operates. While removing Amazon and Google wouldn't take down the *entire* internet, you'd disrupt large parts of it, and much of what these companies do couldn't be quickly duplicated by anyone else. Yes, both these companies have multiple sites for their cloud-hosting machines around the world, and yes, they both try to keep them as secret as possible, but this is still a weak point—especially since that AWS outage we opened this section with occurred in just a single data center.

So: Should you just track down some Amazon and Google data centers and start digging till you cut a bunch of important-looking wires all at the same time? Honestly, you *could*. And it *would* disrupt a large part of the internet for a bunch of people . . .

... but only briefly.

Both of these companies have a huge financial incentive to keep their data centers operative, which means any catastrophic disruption is an all-hands-on-deck, fix-it-at-any-cost, why-did-you-turn-off-the-money-hose-turn-it-back-on-this-instant event. You're going up against some of the largest and richest companies in the world and in likelihood will achieve, at most, a brief and pyrrhic victory: that AWS outage we've been examining lasted only a few hours until it was repaired. Besides, these locations are hardened and secured, and that's only going to stymie you further. You *could* try a more vulnerable target—internet traffic between Europe and the Americas travels on only a few underwater cables across the ocean, and *nobody can defend the ocean*—but even if you discover the exact locations of these cables, the richest and most powerful companies on Earth are just as strongly motivated to fix them as fast as possible, and for the same reasons: they're really, really profitable.

In 2008, cut cables in the Mediterranean Sea instantly curtailed internet access for millions of people in Egypt, Saudi Arabia, India, Qatar, and many other countries. Backup routes were quickly overloaded, reducing Egyptian internet capacity to just 20% of normal—and reducing the Egyptian communications ministry to begging people to stop all the downloading so there'd be more bandwidth left for crucial business needs. This dramatic attack on the critical infrastructure of several nations wasn't pulled off by hackers, elite martini-sipping spies, or even terrorists: it all happened because *some guys didn't want to float away while they fished*. A ship's anchor, inadvertently dragged across the underwater cables, was all it took.* But while cutting these cables is easy—you can do it by accident, after all!—fixing them isn't impossible either: initial repairs here took less than a week.

While capitalism *has* centralized the internet, it's also deeply incentivized those profiting from that centralization to keep their servers up no matter what. Outages don't last long and even unprotected infrastructure

---

*Anchor-based accidental attacks on underwater internet infrastructure are pretty routine: there were several other instances in 2008 alone, they've happened before, and they've continued since. Besides anchors, undersea internet cables have also been damaged by things as pedestrian as the tides and as awesome as attacking sharks.

deep in international waters gets repaired quickly, because every second of downtime means someone very rich and powerful is losing money. Even at your most successful, you might *annoy* these large corporations, but you'll never strike fear into their hearts, nor shall they tremble before you. And worse, your attempts to make them do so will only irritate and inconvenience tons of innocent bystanders who simply wanted to upload pictures of cats and then download other people's pictures of different cats!

There is an idiom involving both hoisting and petards that may be of some use here.

Maybe it's worth taking a step back, regrouping, and looking at this whole "internet" thing again.

Sure, there are a lot of bad things about the internet, but there are also tons of upsides, right? Most of us have had, and probably still maintain, very important relationships with people we would've never met if not for the internet. It's full of incredible facts, hilarious and charming webcomics about talking dinosaurs, videos that entertain for a few minutes before you never think about them again, and, if you've jumped ahead and successfully accomplished the first scheme in Chapter 9, it's also where your eternal villainous manifesto lives. During the COVID-19 pandemic, things would've been indescribably worse all around if we couldn't videoconference with loved ones, and if no one could work from home. And yes, while social media has turned many of our brains to mush, there's no reason that it has to be that way forever. There's no law that says the largest social media platforms *have* to tolerate hate on their platforms, facilitate genocide in Myanmar, or design algorithms to automatically put the most outrageous instances of hate in front of a large percentage of the population because it makes them stay on the sites longer and is therefore profitable. Surely some or all of this could be fixed, and surely by working together, we could, one day, build an internet that's wild and open and free *and* nontoxic *and* nongenocidal *and* good for our mental health.

So that's one argument, founded in optimism, for saving the internet.

Our second and much more *pragmatic* argument is that the following villainous scheme, which depends on the internet being up and available, really will let you take over the world . . . or at least a fair-sized chunk of it.

YOUR PLAN

It turns out there *are* some advantages to a globally connected system of machines that will do whatever they're told to do, and one of the most powerful ways you can exploit said machines is by leaving the internet up and then using it to *steal an election*. Yes, taking over an existing country doesn't give you absolute power (see Chapter 2 for the reasons why), but it does let you get as close as possible to that state without the hassle of having to start your own country. There are tons of perfectly good nations sitting around already—you might as well use 'em!

First, we need to recognize that elections are almost unique in the challenge they present. In a functioning democracy, an election must meet the following five criteria:

1. It must be **open**, since there should be no barriers to eligible people voting.

2. It must be **secure**, since we need to ensure that ineligible people (non-citizens, for example) aren't voting, or that eligible people aren't voting more than once.

3. But at the same time, it must be **anonymous**, because if a citizen fears reprisal based on how they vote, then they by definition can't make a free choice. And going the other way, the impossibility of *proving* how anyone voted naturally limits the lucrativeness of anyone selling their vote to someone else, which is good for the overall health of any democracy.

4. It must be **transparent**, because it's critically important not just that the election is reliable, but that people *believe* it's reliable. The only way to get that is by being fully transparent with how votes are cast, gathered, and counted.

5. And finally, it must be **accurate**, because you only get one shot at an election, and there aren't any do-overs, take-backs, or oopsie-doopsies. Any attempt to annul the results of an election in order to try again will cause massive unrest, as anyone who appears to have won that first election will rightfully feel that they're being robbed.

Already we have conflicting standards: we want things to be secure so people can vote only once, but we also need them to be anonymous so we can never tell how anyone in particular voted. It's a hard set of constraints, but voting on paper actually solves them nicely! Poll workers need only to check identification to ensure that just citizens are voting (**open:** check!), and that they're voting just once at the only poll they're eligible to vote at (**secure:** check!). But once a voter is marked as attending, given their identical and indistinguishable ballot, and admitted to the voting booth, there's no way to connect their ballot to their identity, thereby securing **anonymity**. And it's **transparent**, because while the marking of the ballot is done in private, it's then deposited into a publicly visible ballot box in front of observers. When the election is over, those ballots are also counted publicly, ideally under the watchful eyes of representatives of everyone running in the election. This allows the results to be both provably **accurate** and verifiable, because any dispute about the results can be settled by a recount: gather up the paper ballots, count them up again, and you're done.

But look at the downsides of paper ballots! It takes *so much time* to physically—by hand! like how cavemen would do it!—count up votes. It leads to arguments by poll workers and election observers about hanging chads and signature matches and whether *this* tick is close enough to the ballot area to count when *that* other tick wasn't. Plus, it's real inconvenient to have to tromp down to the poll booth, *especially* when your country doesn't make elections a national holiday! It costs the whole nation time and money to take time off work to vote, which means we could save tons of both by simply doing it online. We all know it's not the 1800s anymore, so why are we still *voting* like it is? Every day we bank online, shop online, file our taxes online, vote in polls to help our favorite stars of reality TV online, and share our choicest nudes online!* Surely adding voting to that list isn't *that* hard. And if it *is* somehow magically impossible to build a simple voting app, can we at least vote on *computers* at polling stations? At least then we could have the results in seconds instead of hours, and recounts could be processed in an instant!

As a supervillain looking to take over a democracy, you *absolutely* want to support and spread the sentiments in that last paragraph. The truth—

---

*Maybe not *every* day. On the good days though.

which any computer scientist not currently trying to sell you on an electronic voting system will tell you—is that you shouldn't put absolute trust in *any* computer, and voting online for anything that matters is *wildly* reckless and dangerous. That's a huge claim, and it's one that might provoke instant skepticism—but that's great for our purposes. The more people who think voting online or on a computer is safe, easy, and convenient, the easier it'll be for *you* to steal their country out from under them.

So! Let's get to it!

How to Hack Computers for Fun and Profit

One possible option for implementing this villainous scheme.

To pull this off, we need to first look a little more closely at how computers actually function. You probably know that computers run on binary: 1s and 0s. You probably also know that manually writing anything in binary is hard—so hard, in fact, that basically nobody wants to do it. Even if you succeed, what you're producing is just a bunch of numbers, something that's very hard for anyone—including yourself in a few days, once you forget what you wrote—to figure out what your code does. Quick, what does this program do?

*Possibly a computer program in binary, possibly the result of me tapping out
the melody of the saxophone solo in George Michael's "Careless Whisper" with
one finger on the 1 key and another on the 0 key.*

The honest answer is "that could actually do anything, depending on how it was interpreted—it depends on what computer it's run on, but even so it'll be a huge pain to figure it out."

To avoid this problem, we computer scientists invented something called "assembly languages." These languages are directly based on—and tied to—the hardware and binary code that computers run on, but they're made more understandable by being closer to languages we humans actually speak. In assembly language, adding 10 and 20 might look a little something like this:

```
MOV 10,REG1
MOV 20,REG2
ADD REG1,REG2,MEM3
MOV MEM3,SCR
```

*Adding 10 and 20 in an imaginary but representative assembly language. I used an
imaginary language here because it's clearer, and also because even though I have
two degrees in the computer sciences, I still thought I might make a mistake if I used a
real one. But I definitely didn't, because I hereby formally define this language I
invented as "whatever makes the above code spit out the number 30."*

As you can see, there are obvious downsides here. Assembly languages are painful, and they're tedious, and the enormous amount of clerical detail you need to keep track of (memory locations, whether or not your registers are filled, and so on) makes it very easy to make a mistake. You need to be intimately familiar with your computer's hardware to write in assembly, and since different processors all have slightly different architecture, there's always more to learn. But at least when compared to binary code, what's supposed to happen here is much clearer than it would be if you were trying to read an endless stream of 1s and 0s!

The next step up is to abstract away the hardware, so you don't need to know the location (or even existence!) of small pieces of computer hardware like "adders" and "registers." Languages that do this are called high-level programming languages, and they're a much more intuitive way to create software. Using them you can write something very close to "add 10 and 20 together and print that result to the screen," and it'll *just work*. And the reason it just works is because when you're ready, you run another program—called a "compiler"—to translate your high-level code down into the most efficient binary code for whatever computer you target. Printing the sum of 10 and 20 in a high-level language is as simple as writing this:

print 10+20

Adding 10 and 20 in a higher-level programming language.
The result is 30. I promise it's 30.

The Joy of Compiling High-Level Languages

Anyone who has developed software in high-level languages before is probably already halfway through composing an angry email in their head to me addressing that "compilers just work" line, because they know how subtle bugs in programming can be. Early on in my studies, I once spent four hours trying to figure out why a program I'd written would fail to compile, only to finally notice that while every line in the language I was using had to end in a semicolon, on one line I'd accidentally let go of the "shift" key on my keyboard too late and typed a colon instead. That substitution—the result of a possibly microsecond-sized change in how long I pressed the "shift" key and manifesting as a *single pixel's worth of difference on my screen*—burned up my whole afternoon. Anyway, computers are great!

Today there are hundreds of high-level programming languages—each with their own philosophies, specialties, and use cases—but they all share the same goal: to make computer code easier for humans to read and write, which in turn makes computer software easier to build, understand, and maintain. High-level languages have unlocked the potential of computers, allowing us to build larger, more complicated, and more beautiful software. They free us humans to operate at the level of *ideas*, while our compilers automatically handle the complicated, monotonous, nitty-gritty machine-level details of *implementation*. With the exception of hobbyists and some relatively small sections of specialized software,* today all software is written in high-level languages.

That's precisely what makes them *so* vulnerable to the evil machination you're about to pull off.

---

*A few decades ago, when compilers couldn't always write code as efficiently as a human would, software developers would sometimes drop down into assembly when writing critical segments of their program, in order to wring every last bit of performance out of their machines. But these days that's rare: not only are computers faster, but compilers now generate programs that are more efficient than anything an average human developer could come up with. It makes sense: once a clever human builds a better compiler, that optimization can be used in similar compilers moving forward, causing them to become gradually more efficient over time.

Low-Down Dirty Tricks for High-Level Languages

The weakness is hidden there in plain sight: when you write code in *any* high-level programming language, you're trusting the compiler to *accurately* transform what you wrote into binary code. To mess with someone's programming, all you need to do is mess with their compiler. There's an obvious way to do this, and there's an insidious and nigh-undetectable way.

Let's do the obvious way first.

Here's what normally happens when you compile an application written in a high-level language:

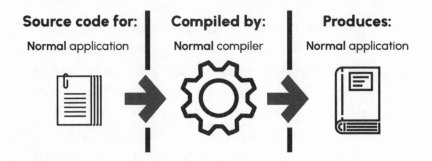

Source code for:
Normal application

Compiled by:
Normal compiler

Produces:
Normal application

Let's assume you've broken into your victim's computer over the internet, secretly altering their compiler so that when it compiles code with the "print" command, it'll add 1 to any number it displays. Now, when anyone uses it to compile an application, the process will go like *this* instead:

Source code for:
Normal application

Compiled by:
Villainous compiler

Produces:
Villainous application

Suddenly—even though coders hadn't made any errors in their programs—anything they build using this altered compiler will happily inform the world that 10+20 equals 31. Applications will now act in ways that were never intended, and poor victimized programmers won't be able to find and fix this "10+20=31" glitch by looking at their own source code— the first place they'd check—because *that's not where this change lives*. It's hidden in the source code of the compiler.

A successful hack.

Of course, these people will realize something screwy's going on pretty quickly, because their programs are malfunctioning in an obvious and detectable way: every single number will be off by one. But what if you got a little more subtle? What if instead of messing with the "print" command, you changed it so that whenever their compiler detected code involving passwords, it made it so the password "ryaniscool" *also* worked? (You could use your own name here, but I will now discreetly suggest that using mine might throw the authorities off your trail for a while: *my gift to you.*)

Do that, and you'll have what's called a "back door" into every computer program anyone builds with your compiler: the metaphor is that your victims can lock and alarm and secure their front doors all they want, but it doesn't matter because you'll have a secret door 'round back that nobody knows about. No matter how talented the developers are and how secure their password code is, anything built using your villainous compiler is also going to let in people using "ryaniscool" too—and unless these developers somehow randomly happen to try that password, *they'll never even know.*

Here you're probably thinking, "But you already said compilers are computer programs like any other, which means anyone could look at a compiler's source code and find my changes. Once they remove the part that makes 'ryaniscool' a valid password, that's my scheme foiled."

And you're right: anyone could do that. But you said it yourself: compilers are computer programs like any other. That means they *themselves* are compiled.

Here comes the insidious part.

Step 1

We'll begin by running that same scam we walked through in that last example—adding the "ryaniscool" backdoor to a compiler—but this time we'll go through it in a little more detail. Before you got your hands on it, the unaltered process for compiling a compiler looked like this:

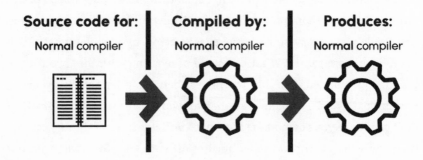

Source code for:
Normal compiler

Compiled by:
Normal compiler

Produces:
Normal compiler

Like last time, you'll alter the compiler's source code so that every application it builds treats "ryaniscool" as a valid password. So now compiling the compiler looks like this:

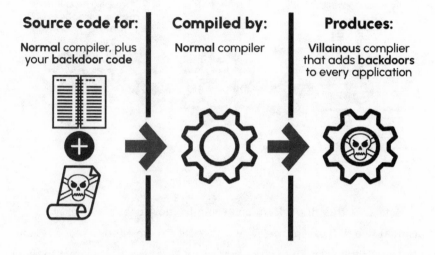

Source code for:
Normal compiler, plus your **backdoor code**

Compiled by:
Normal compiler

Produces:
Villainous complier that adds **backdoors** to every application

And just like last time, you've now successfully built a villainous compiler, but you'll still be caught the second someone looks at its source code, where all your changes lie in plain sight. Good thing we're not done yet!

Step 2

You're going to add even more malicious code to this compiler: it'll now detect when it's compiling *itself*. Whenever that happens, your compiler will now *reintroduce* the code you wrote that instructs compilers to build your backdoor into every application it touches. So now the process looks like this:

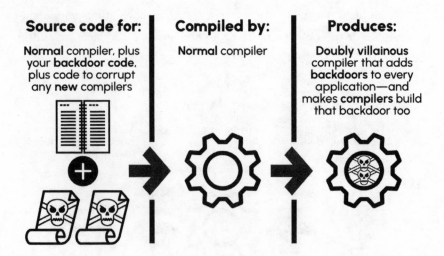

Source code for:	Compiled by:	Produces:
Normal compiler, plus your **backdoor code**, plus code to corrupt any **new** compilers	**Normal** compiler	**Doubly villainous** compiler that adds **backdoors** to every application—and makes **compilers** build that backdoor too

Note that this new villainous compiler now behaves exactly as the original one did, except under two specific circumstances. When it compiles applications, it'll introduce that secret "ryaniscool" backdoor to them as before:

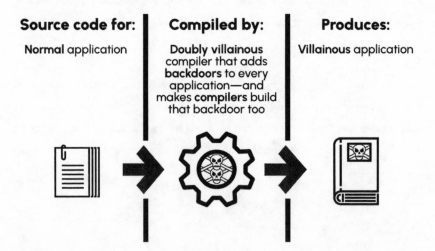

Source code for:	Compiled by:	Produces:
Normal application	**Doubly villainous** compiler that adds **backdoors** to every application—and makes **compilers** build that backdoor too	**Villainous** application

And when it's run on *itself*, it'll build a new compiler that also adds your backdoor to every application *it* builds too. Your changes are now impossible to get rid of, *even if someone builds a new compiler from tested, trusted, and unaltered source code:*

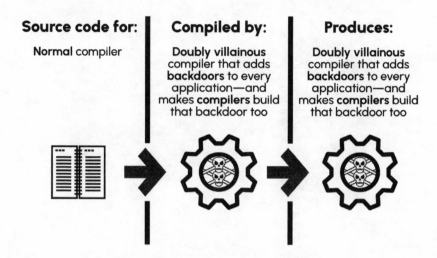

Source code for:	Compiled by:	Produces:
Normal compiler	**Doubly villainous** compiler that adds **backdoors** to every application—and makes **compilers** build that backdoor too	**Doubly villainous** compiler that adds **backdoors** to every application—and makes **compilers** build that backdoor too

Step 3

This heist is just about completed. All you have to do is remove your malicious instructions from the compiler's source code, and *you've just perfectly covered your tracks.* This compiler now behaves (as far as anyone can tell) exactly as it's supposed to, happily building programs that appear for all the world to be unaltered. But it's actually hopelessly compromised: any application it builds is now to *your* specifications, and any new compiler it builds will secretly also be constructed with your instructions baked in.

And all this happens without a *single* line of evidence being left in any source code.

*An even **more** successful hack.*

And that's it! You're done. Change the password example we've been using to "count votes for yourself more than votes for anyone else," and you can consider this heist *completed*.

Oh! And congratulations on your performance at the digital ballot box, *Madame President.*

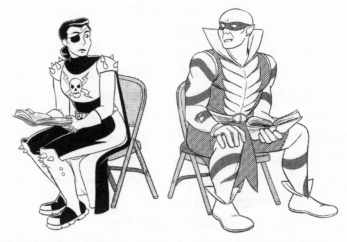

If there is another supervillain competing on the ballot you're interested in, simply ensure that you're the faster reader, thereby finishing this chapter before they can.

Nice.

It is nice! It's very nice. And while you might think your victims could just walk through the binary code of your program itself to check it, this is a task that starts out hard and becomes only more difficult as programs gain complexity. For comparison, the complete works of William Shakespeare—which are at least written in English!—come in at under 6 megabytes (MB), and it would take you several days to read the over 50 million bits of data that represents. The Firefox browser alone requires over 200MB just to install, and that's just one program on your computer. There is likely not a human alive who has read all 200MB of that binary code. It's not even written in a language designed for humans to read! And yes, if someone suspected you'd compromised their compiler (which, if you've done your job well, they never will), they could at least compare it to a known-good compiler and examine the sections of binary code that have changed. But putting aside the question of where they'd get a known-good compiler from, this wouldn't tell them what it's doing differently, just that it's changed. To figure out what it's doing differently requires stepping through the binary code and reverse-engineering it from scratch.

The best (or worst, depending on your perspective) part of this scheme is that none of this is secret and that this weakness has been known for decades. In 1984, Ken Thompson—the man who designed and implemented Unix, the progenitor of the operating systems most computers and phones run on—produced a paper called "Reflections on Trusting Trust," which reached the same conclusion: "The moral is obvious," he wrote. "You can't trust code that you did not totally create yourself. (Especially code from companies that employ people like me.) No amount of source-level verification or scrutiny will protect you . . ." By "totally create yourself," Thompson doesn't just mean a program that you wrote, but one you wrote the entire stack for: everything down to the compiler. Very few people have the time, skill, and money to build a computer from the ground up, including all the software on it. This would seem to be a bullet in the head for trusting

not to be trusted

The V1L-N80 is more than just a computer; it's a threat. This versatile machine stores all of your personal information, plus it's better at math than you. Look at its inscrutable little screen—that's the face of a criminal mastermind.

You've been warned!

Victory.

computers with anything. And honestly, if you've ever felt like laughing villainously, *now would be the time.*

But despite this, we still trust computers with all sorts of things! So, what gives? Why are we willingly using these nightmare machines?

It all comes back to those requirements of voting: open, secure, anonymous, transparent, and accurate. Most things we use computers for—heck, most things we encounter in real life—either don't share *all* those properties, or if they do, their stakes aren't nearly as high. People use their credit cards online with confidence, but that's only because the credit card companies are willing to cover the fraud that happens as a business expense: it's worth it to keep the money (and income) flowing freely. Banks make the same financial calculation when they promise transfers using their websites are secure: sometimes criminals *do* get away with money, but in those cases the banks cover the loss because the value of everyone trusting their website is worth more than what the fraud costs. Voting in online polls works because the stakes are lower—if the results are altered, discarded, or ignored, the fate of a democracy doesn't hang in the balance, because *it's just some stupid online poll.* And when nudes leak, it's generally because they were shared with someone who wasn't worthy of them and who leaked them himself, not because the computers holding them were hacked.* The reality is, computers are both fun and convenient, and there aren't that many cases where you really *do* need absolute trust in what they do.

But elections are not one of those cases.

Elections are a case where a hack can have a huge and irrevocable effect, because we can't easily correct elections after they've happened. And they're by their very nature extremely public, with when and where voting's going to take place being announced well in advance. Worse still, there's a clear, desirable, incentivizing, and potentially quite-profitable payoff to altering

---

*But of course, the latter has happened too, most notably with a large set of celebrities' private photos that were acquired by compromising Apple's iCloud photo-hosting service in 2014. At the time, Apple's service allowed an unlimited number of attempts to guess a password, which meant malicious hackers could automate the process, trying over and over until they gained access.

the outcome of an election—and these incentives apply especially to other nations, who could more easily marshal the resources required to pull off a heist of this magnitude: a heist of *democracy itself*. And this hack can be accomplished by gaining physical access to the machines, by breaking into them remotely over the internet, or even by having just a single compromised USB drive inserted into a machine on a private *internal* network.

Sounds like tinfoil-hat conspiracy-theory stuff? Alarming but hypothetical attacks that are far too complex to ever be pulled off in real life? Turns out, very similar attacks have already taken place. And they were deployed not against publicly accessible voting machines, but against something even more hardened and secure: *uranium-enriching nuclear centrifuges.*

How to Sabotage the Nuclear Programs of Foreign Nations for Fun and Profit

In 2010, a malicious piece of software that spread itself through then-unknown vulnerabilities in Microsoft Windows was discovered: the Stuxnet worm. What made Stuxnet remarkable was how specific it was: it did nothing unless it found itself infecting specific Siemens industrial control systems, which were a hardware platform used, not coincidentally, for uranium enrichment in Iran. And Stuxnet was stealthy too: not only did it avoid damaging any unrelated systems it infected, it was also programmed to erase itself entirely on June 24, 2012. But before that date, when it found itself on the right hardware, Stuxnet did something truly devious: randomly, and only occasionally, it would change the directives of the centrifuges it was attached to. Stuxnet made them spin so quickly that they would tear themselves apart, while *simultaneously* ensuring that the attached computers would report that nothing was going wrong. It would be an infuriating, frustrating, and impossible-to-diagnose bug for the Iranians: all reports and recordings would show everything was working perfectly, except for the fact that sometimes, for no apparent reason, some very expensive hardware would simply self-destruct.

And all the evidence we have shows that Stuxnet worked: while the Iranian authorities have never conceded the cause, in 2009 and 2010, accidents and malfunctions at their uranium-enrichment facilities spiked far above normal levels. United Nations inspectors watched as around 2,000 of the 9,000 centrifuges at the enrichment facility in Natanz were shut down, dismantled, and eventually replaced. Stuxnet was controlled by and eventually spread to the internet—where it contacted specific servers in Denmark and Malaysia to receive commands and updates to its code—but it also spread if an infected USB stick were inserted into a centrifuge-controlling computer, or into any other Windows computer linked on the same ostensibly protected *internal* network.

Since the computers controlling Iranian uranium enrichment weren't connected to the internet for security reasons—Iranian officials knew just as well as you now do that this was a terrible idea—an infected USB stick was likely how the infection made it inside the facility. And it didn't even require elite superspies to break inside the Iranian facility to make it happen: it was likely done by some hapless worker who either found or was given the USB key, and who may not have even realized what they were doing.* Stuxnet's earliest infections were within five Iranian companies that worked with the nuclear program as contractors, from which it could've spread, system to system, until it had finally reached its target.

The authors of the worm have never been publicly identified, but it's believed to be a cyberweapon built through the joint effort of both the U.S. and Israeli governments and then specifically deployed against the Iranian nuclear effort, since over half of all infections were in Iran. Stuxnet's complexity and sophistication suggests it took as many as 30 coders upward of six months to build, with later analysis by security researchers at Kaspersky Lab concluding "this is a one-of-a-kind, sophisticated malware attack backed by a well-funded, highly skilled attack team with intimate knowledge of SCADA [the Supervisory Control and Data Acquisition part of uranium-refining] technology. We believe this type of attack could only be conducted with nation-state support and backing."

In other words, a bunch of sovereign nations have already decided that it'd be in their best interests to meddle with computers of another sovereign nation in order to further their own interests. Despite the fact that the Iranian machines weren't hooked up to the internet, they were still attacked remotely using the internet, in a way designed to be undetectable. Excluding the initial infected USB stick that had to get into the right hands in Iran, the rest of this attack—its development, its management, its oversight, and its updates—were all done from the security and safety of *another country*, with thousands of kilometers between the hackers and their intended targets. And nobody has ever been prosecuted for it! The only reason we know about Stuxnet is because the software later infected more public machines on the internet before that 2012 self-destruct date.

---

*In 2019, it was reported in *The Jerusalem Post* that a mole working for the Danish intelligence agency AIVD was recruited to bring that first infected USB key to Iran, where they either inserted it themselves into target computers or manipulated someone else into doing it. It's possible that the original USB key wasn't even infected when it entered Iran: sometimes the easiest way to get digital contraband across a border, once you're safely and nonsuspiciously inside the country, is to simply download it from a secret internet server you've built beforehand.

But then again, why *would* anyone be prosecuted? The developers were outside Iran's jurisdiction, no country has ever admitted to being responsible for the software, and many are more than likely *thankful* to the developers who built it. The Stuxnet sabotage bought time for the international community and put pressure on the Iranian government, and it may even have been a factor in Iran signing a nuclear deal in 2015, in which it agreed to reduce its uranium stockpile and shut down most of its centrifuges in exchange for international sanctions being lifted. This sort of targeted, elite-tier hacking of mission-critical computers, secretly compromising them to further your own interests at the expense of someone else's—it's not just theoretical, it's *already happened.*

It all leads to one conclusion: computers aren't safe, which means computer voting isn't safe either, and the internet lets you exploit these machines from the other side of the world. The *only* safe-ish way to vote with a computer is in person, using a system in which a paper ballot is printed in sight of the voter, approved, and then stored in a publicly visible and secure ballot box: a voter-verified physical ballot. That way, if someone thinks the computer systems were compromised—if there's any reason at all to suspect someone added the votes improperly—there's a paper trail. The computer is used as a convenience, and the real vote, the real power, still lies in the paper ballot. Without that paper trail, the public is left trusting the computer.

And as you now know, nobody should ever trust a computer.

But thankfully, most people aren't familiar with Stuxnet or the nature of compilers, and they all really, really want to use computers to vote! So by encouraging the public in its desire for voting in elections on computers or over the internet or via some dinky free app while waiting in line for coffee, you make it that much easier to undermine or even take over their government from the comfort and safety of your own home, secret base, or competing nation.

Remember me when you do. I am available for any and all sufficiently prestigious cabinet positions.

SPECIAL INTERACTIVE SUPERHERO Q+A SECTION

Go for it!

That's more of a comment than a question, but sure: in the "change the value of what 10+20 adds up to" example, that'd obviously be easy to test for and catch any changes. But even if someone thought to test that in the first place—and why would they?—that still doesn't solve the core problem. If you're smart, your malicious code could detect when it's being tested and behave normally, only changing things when people aren't looking.

Sounds like sci-fi, right?

Well, this has already been done too: in the 2015 Volkswagen emissions scandal, the car's onboard computers detected when their emissions were being tested and ran in a low-power, environmentally friendly mode, switching back to their high-power polluting mode once the test was over. They were explicitly designed to be on their best behavior when being scrutinized, and then, when they weren't, to dump as much as *40 times* the legal amount of pollutants into the atmosphere. This scandal has cost Volkswagen over $37 *billion USD* in fines and settlements so far, with another couple billion still to come.

The only reason Volkswagen would do this in the first place is because they thought it'd be a profitable thing to try, and because they probably believed they wouldn't get caught. The same incentives apply even more strongly to stealing an election.

In an absolute sense: yes. Nobody should have 100% faith in any computer system. But that's obviously not practical, and in most cases, they don't *need* to have 100% faith in a computer. Remember: in the vast majority of the things we use computers for, a malfunction—whether malicious or accidental—can be covered for. The damage gets fixed, and we all move on with our lives. In 2019, when my bank got hacked and my personal information was stolen, their official response was to send me a brief letter in the mail. In it, they kinda but not quite apologized, assured me they were doing all they could to prevent this from happening in the future, suggested it maybe wasn't such a big deal anyway, but *just in case* they concluded by giving me a code valid for five free years of credit monitoring services. I accepted for three reasons: because no money had been stolen, because the culprits would likely never be found so this was as close to justice as I'd ever get, and because the *other* bank I use had sent me a very similar letter just one week earlier after suffering their own unrelated hack, but *those* jabronies only offered one single year of credit monitoring.

Even in personal finance, when compared to the fate of a democracy, we can see that the stakes are lower, and we're willing and able to compromise. But (thankfully for us supervillains), there's no way to correct a broken election after the fact.

If you're not familiar with them, blockchains (used for mining, buying, and selling cryptocurrency like Bitcoin) are effectively a distributed database wherein changes, once accepted, are permanent and public—and any change you make is tied to your blockchain ID. If votes were recorded in the blockchain, then the second anyone managed to connect your blockchain ID to your real name, the entire history of every election you ever voted in is instantly and irrevocably public. Just the threat of this breaks anonymity, which means blockchain voting does not support democracy, nor does it solve the exploit we're using here. Oh well!

Absolutely. But paper ballots have a few huge advantages: their downsides are intuitive and well understood, and their vulnerabilities are limited to those who have physical access.

If I want to mess with a paper ballot election, I need to either steal, alter, or stuff ballots—which means that whichever option I choose, I'm gonna need physical access to that ballot box. And that limits the amount of damage that one bad actor can do. The issue is one of proportion: with paper ballots, one person might, at best, interfere with a single polling station, or maybe a couple of them if they could really boot it around town on election night. With computer voting, they can interfere with all of them, at the same time, from literally anywhere on the planet.

On top of all this, there's the simple fact that programmers aren't perfect. Even if the attack I've described here isn't used, that doesn't mean your computer voting system is secure. Heck, Google—who is generally understood to hire some very smart people—has a bounty system in which it will pay anyone cash money if you help it find bugs in its own software, because it can't *guarantee* it hasn't screwed up somewhere. It turns out that paying ethical white-hat hackers to tell you about your mistakes is cheaper, in the long run, than cleaning up the mess caused by unethical black-hat hackers exploiting those same weaknesses for profit. The Stuxnet worm we examined earlier worked by using bugs in Windows that hadn't been fixed, because *not even Microsoft knew about them.*

Software programming is hard. Computers are hard. And even a brilliant, well-intentioned software developer can make a single mistake that opens an entire system to intrusion. For an example of that, you don't have to look further than one of the most popular pieces of free software used to handle encryption for websites: OpenSSL.

In 2011, a single line of OpenSSL code introduced what would eventually be known as the Heartbleed bug. It was a simple programming oversight that allowed attackers to read *anything* stored in the memory of a machine running OpenSSL, including passwords. There the bug remained in plain sight in the OpenSSL codebase: open-source software that in theory anyone on the planet could've downloaded, examined, and corrected. It was only in 2014 that it was finally noticed (at least by someone who disclosed it, rather than exploiting it), at which point it was quickly corrected. But by that point, 17% of the secure servers on the internet were vulnerable, and they would remain vulnerable until each of them was updated. In the immediate aftermath, operating systems had to be upgraded, websites who couldn't tell if they *had* been hacked but knew they *may* have been asked users to reset their passwords, and the Canadian federal government was forced to take their revenue agency's website offline after hundreds of social insurance numbers were stolen by people exploiting the now-public bug before they could patch it. *Forbes* described it as "the worst vulnerability found (at least in terms of its potential impact) since commercial traffic began to flow on the Internet."

Heartbleed was coded by *accident*.

Like the Thomas Midgley we met way back in the introduction to this book—remember him? *He managed to threaten all human life on the planet twice by mistake?*—you have to imagine what a sufficiently motivated software developer could do if they were really trying.*

---

*Past errors in computer voting software (among those that were caught, at least) include a 2003 election in Fairfax County, Virginia (in which the machines altered one out of every 100 votes for a specific candidate to be in favor of her competitor instead), a 2003 election in Boone County, Iowa (where operator error allowed the machines to report 140,000 votes cast in a community with 50,000 people), and in the 2000 federal election in Florida (where a computer voting system reported presidential candidate Al Gore as having received *negative* 16,022 votes).

It's true: some countries use computers to collect votes, and some even allow online voting. But that doesn't prove that it's safe, just that it's *popular*—at least among people who are already in power and who would presumably like to keep it that way. Some countries have hedged their bets by putting restrictions on computer voting: in Switzerland, only 10% of the electorate may vote online. (This is presumably to limit its effect on elections, though of course a close election would allow corrupted machines to affect the result.) France allowed internet voting in 2003 but suspended it in 2017 due to security concerns. Germany began a trial of computer voting in the early 2000s but ended the trials in 2009, declaring that the inability of the general public to understand the machine's source code made them unconstitutional.* And in 2019, even Switzerland saw calls to suspend their internet voting after a flaw was discovered that allowed votes to be manipulated without that manipulation being detected.

But don't worry, supervillains: there are still places like Brazil (with computer voting in place since 1996), India (with electronic voting since 1998, though after complaints of fraud in 2014, some paper trails were introduced), and the United States (with some states forcing citizens to vote on a computer that doesn't generate a paper trail, whose source code they cannot examine, and which was in any case likely built using a compiler the machine's designers didn't write themselves.†) If you're looking for an entry-level practice country to take over first, these have already done a *lot* of the prep work for you.

THE DOWNSIDES

The downsides of pulling this off are obvious and raise the question: Should you *really* use the internet to hack computers, steal an election, and gain control of a country? It seems like there might be some pretty significant downsides to doing that—if not for you, then at least for democracy as a whole. And yes, even though we are all supervillains here, I am going to go out on a limb and say—perhaps uncharacteristically—that interfering with democracy is bad, actually, and the ends can't justify the means. But

---

*Though as we know, even looking at the source code wouldn't necessarily be enough to find malicious code.

†It's worth noting that in the 2020 American federal election, the states that outgoing president Trump claimed had "rigged" elections—including Georgia—had spent millions of dollars replacing their old no-paper-trail computer voting machines with ones that do actually create a countable, voter-verified physical ballot. However, the eight states in America that still use paper-trail-free computer voting machines (Indiana, Kansas, Kentucky, Louisiana, Mississippi, New Jersey, Texas, and Tennessee) *weren't* battleground states, and all except New Jersey went comfortably to Trump. For some reason, those states weren't the focus of the president's conspiracy theories.

my telling you precisely how to steal an election and then getting all holier-than-thou about it is in service of a *larger* villainy.

Yes, you now know how it could be done, and yes, you now know to never trust computer voting for anything that matters. But more importantly, you also know to demand more from your elected officials—and your elections—than the vulnerable and insecure farce that is voting on computers and/or the internet. This allows you to be a thorn—or better yet, a *supervillain*—in their sides, demanding they do better until they're forced to acquiesce to your demands for actual physical paper ballots. It's like we saw in Chapter 2: the way you force people in power to do what you want is by ensuring that they fear you.

And at the end of the day, there's nothing more supervillainous than taking steps to ensure that you're feared by *the most powerful people on the planet*.

POSSIBLE REPERCUSSIONS IF YOU'RE CAUGHT

If you *do* decide to go ahead and steal an election, blithely ignoring that brief three-paragraph and perhaps-legally-mandated "okay, maybe don't pull off this heist" advice, but heeding my extremely detailed, chapter-length "now just between us friends, here's *precisely* how you'd pull off this heist" advice, you'll be playing a dangerous game. Countries really don't like it when you meddle in their elections . . . buuuut, as we've already seen, so long as whoever's facilitating that meddling stays outside the country whose election they're undermining, there's not much anyone can do, and there are certainly no legal consequences they can impose. And once you're in charge of a country, you generally have a lot of influence in deciding which crimes get prosecuted under the full extent of the law and which get quietly swept aside forever.

It's important to note that with this heist, the repercussions depend not only on *if* you're caught but *when* you're caught. If people believe an election was fair, there's not a ton they can do to correct things two terms later when they find out it wasn't. Remember that Volkswagen emissions scandal? They got caught in late 2015, but Volkswagen had been cheating on emissions tests for *six years* before then. In America, those six years would be enough for you to win not only your first term as president but to also be halfway through your second term before anyone noticed something was wrong. Heck, even if your meddling is revealed only a few months after the election, that might still be enough time for a majority of people to have decided you've won and to not want to hear otherwise. There's a huge cost—social, financial, societal, political—to overturning elections, and the longer you get away with it, the greater those costs become for everyone else. If you're lucky, you might quickly reach a point where the cost of undoing an election is so high that a sufficient number of people no longer wish to pay it. When that happens, it becomes easier and more profitable for them to sit back, relax, and just kinda see what you're gonna do with the place now that you're in charge. This is, it's worth noting, also the point where you do really and truly win.

And you didn't even have to blow up the internet to do it.

PART THREE

THE UNPUNISHED CRIME IS NEVER REGRETTED

HOW TO BECOME IMMORTAL AND LITERALLY LIVE FOREVER

I think life is a colossal tragedy. Everything goes well—people enjoy life, study, get married, get divorced, have aspirations, build careers . . . And then they start rotting alive.

—*Mikhail Batin (2013)*

Let's ignore the people alive today and focus instead on the previous 200,000 years of humans who came before. If we look at them, we find a rare instance where we can generalize across a tremendous swath of humanity, across every civilization that has ever existed, across humans who were brilliant, or indolent, or rich, or poor, across serfs and sovereigns and scientists and shamans, all the way into prehistory, to before prehistory, to the very moment when the first human (however you decide to draw that line) was conceived, and born, and opened their eyes to a pale blue sky as full of potential as they were—we can generalize across all these people who lived in times and places we may only ever understand dimly if at all, and we can still have what we're about to say be fully accurate down to every single man, woman, and child: *everyone who has ever tried to live forever has died in the attempt.*

Without exception, everyone who thought they'd found a path to immortality was absolutely wrong, this idea has a failure rate of 100%, and not even a single human being has managed to live forever in the 13.8 billion years the universe has existed. Not one.

But then again, it's equally true to say that in all those 13.8 billion years, there's never been a human being *quite* like you.

BACKGROUND

Live-long-quick schemes are almost as old as civilization itself. Here are some chronological-order classics:

- the Fountain of Youth, whose waters (when consumed or bathed in) cure illness and make you young again (Greece, 400s BCE)

- the Philosopher's Stone: a substance that transmutes iron into gold and also grants immortality when consumed (Greece, 300s BCE)

- drink enough deadly mercury and you'll become immortal (China, 210 BCE)*

---

*This refers to Qin Shi Huang, who in 221 BCE became the first emperor of a unified China when he was 38 years old: he actually created the title "emperor" for himself. He ruled for 11 years before he died, and in that time he was obsessed with immortality. He sent hundreds of men and women to find the mythical Mount Penglai, where an enchanted fruit that granted immortality was said to grow, and eventually ordered a full nationwide search for any and all elixirs of eternal life. In 2002, pieces of ancient bamboo and wood writing were found in an abandoned well in Hunan Province, and among them were slips containing not only the emperor's orders for this search, but some with responses: the village of Duxiang reported no luck yet in finding any immortality potions, while the village of Langya suggested he try a herb from a nearby mountain. When he died at 49, still in middle age, Qin Shi Huang had likely been consuming large

- replacing the blood of the old with the blood of the young to recover youth (England, 1600s CE)*

- track down the literal tree of life from the biblical garden of Eden (either saved by Noah *on* his ark, used by Noah *for* his ark, or surviving in some descendant somewhere: "a most wholesom, odoriferous, balsamical, and almost immortal Shrub") and use medicine derived from its wood to live forever (Spanish Netherlands, 1600s)

- live without sin and you'll never die (England, 1650)

---

amounts of mercury in the hopes it would make him live forever. Emperor Qin Shi Huang was not the first to die from what's now known as "elixir poisoning," nor would he be the last. In fact, over 700 years later, Chinese alchemists were still describing what we now recognize as the *actual symptoms of heavy metal poisoning* as proof an elixir was working: a text around 500 CE reads, "After taking an elixir, if your face and body itch as though insects were crawling over them, if your hands and feet swell dropsically, if you cannot stand the smell of food and bring it up after you have eaten it, if you feel as though you were going to be sick most of the time, if you experience weakness in the four limbs, if you have to go often to the latrine, or if your head or stomach violently ache—do not be alarmed or disturbed. All these effects are merely proofs that the elixir you are taking is successfully dispelling your latent disorders."

*This literally vampiric idea was popular in England at the time, and Robert Boyle (one of the founders of modern chemistry) took a deep interest in it. When the first successful transfusion between live animals took place in the mid-1600s (it was a gruesome affair, performed without anesthetic, and the whimpering donor animal died during the attempt), Boyle wrote a journal article filled with questions he wanted explored in further experiments: Would a cowardly dog, upon receiving the blood of a fierce one, become ferocious? Would it still know its old tricks and recognize its master? Could a blood transfusion cure disease, and could a dog be kept alive indefinitely via frequent injection of new donated blood? The dog experiment led to attempts to transfuse blood between a lamb and a human, in the hopes of both curing disease and rejuvenating the aged: this didn't happen, and public mockery afterward put an end to English transfusion experiments for decades. But Boyle never got out of the immortality game: the first two things on his list of the 24 technologies he hoped science would one day develop were "The Prolongation of Life" and "The Recovery of Youth, or at least the Marks of it, as new Teeth, and Hair color'd as youth." (These were followed by some recognizable technologies we've since achieved: human flight, "Potent Druggs to alter or Exalt Imagination," and "the practicable and certain way of finding Longitudes," along with several more exotic technologies we haven't *quite* nailed down yet, such as transmuting metals, "The Cure of Wounds at a Distance," and, perhaps most alluringly, "Attaining Gigantick Dimensions.") Boyle never gave up on seeking the Philosopher's Stone, and in 1678, he claimed to have at least performed a successful experiment with its opposite: using a dark reddish powder he called "the anti-elixir," he said he'd turned molten gold into a lesser silvery metal. Anyway, none of this worked; he died in 1691!

- consume pills made of arsenic, mercury, antimony, and other poisons; assume that the fact your skin is now drying up, your hair is falling out, and your nails are falling off is actually proof that the pills are working; and then, if you survive, take the regrowth of brand-new hair, nails, and skin as proof of rejuvenation (Europe, 1660s)

- restore the world to how it had been before the biblical flood, which, it was believed, cracked the hitherto smooth and featureless world open like an egg and shifted it off its axis, and having accomplished that, eat only the nourishing foods that will be restored to it then, foods that are no longer found in our decayed, ruinous, less fertile, and almost-unrecognizable fallen planet, and you'll never die . . . except from accidents (England, 1684)

- lock a rooster alone in a coop for fifteen days but keep him well fed with good wheat, then provoke him into a jealous rage by letting another rooster hang out with six hens and eat his food while he watches, then kill that initial rooster and drench him in his own blood three times, then combine that blood with three drops of oil of ambergris (itself a complicated recipe calling for distilled wine "such that will set gunpowder on fire," and the "Spirit of the Flower of Mercury"), and take a spoonful of the resulting liquid in the morning for fifteen days in a row to restore youthful strength (England, 1722)*

- harness the power of the mind: be cheerful, think only positive thoughts, and cultivate yourself beyond the need for sleep (which is the very image of death), and you'll live forever (England, 1793)

---

*The book this recipe was found in also includes similarly rococo directions for a "universal medicine," as well as the "sovereign essence," which, it is claimed, can "repair the Decays of Old Persons the most exhausted." They're all found in Harcouet de Longeville's charmingly titled *Long Livers: A Curious History of Such Persons of Both Sexes who have liv'd several Ages, and grown Young again: With the rare Secret of Rejuvenescency of Arnoldus de Villa Nova, And a great many approv'd and invaluable Rules to prolong Life: as also How to prepare the Universal Medicine.*

- sharing the blood of the young with the old to recover youth, and while you're at it, sharing the blood of the old with the young to cure tuberculosis (Soviet Russia, 1920s)

- surgically graft slices of baboon testicles into the scrotums of old men to prolong their lives (France, 1920)*

- get a half vasectomy, because then you'll keep the life-giving sperm from that one testicle inside your body, which will then rejuvenate you (Austria, 1920s)†

- dietary supplements and creams that will make you both feel younger and live longer (a multilevel-marketing pyramid scheme near you, current day)

---

*Yes, this is the footnote for people who want to learn more about baboon testicle implants! These operations were performed by a French surgeon named Serge Voronoff, and over the course of the 1920s, thousands of men were treated with this surgery, eventually inducing Voronoff to open up his own monkey farm to keep up with demand. He was inspired by the work of an earlier Mauritian doctor, Charles-Édouard Brown-Séquard, who in 1889 (at age 72) hypothesized that the liquid extracted from ground-up dog and guinea-pig testicles would rejuvenate him, make him hornier, and prolong his life. He prepared the mixture, injected himself with it, and announced in a public lecture that afternoon that he'd had sex with his wife that very morning, felt 30 years younger, and considered the experiment a success. He died within five years of these experiments. Voronoff ran with the idea—which led to the testicular tissue transplants previously mentioned—along with analogous later experiments for the ladies in which he transplanted monkey ovaries into women. (It's worth mentioning that outside of longevity experiments, Voronoff also tried implanting human ovaries in a female monkey and then inseminating it with human sperm. The early 1900s were a wild time in medicine.) Voronoff believed that once science discovered what chemicals testicles excreted, transplants would no longer be necessary and that this would lead to a coming world "populated by a race of supermen" who "should live twice the normal span." By the late 1920s, there was a growing chorus of critics, including an editorial in the American Medical Association's journal that called his procedure "useless and even harmful." With the 1935 discovery and isolation of testosterone, Voronoff hoped he'd be able to silence his doubters, but experiments quickly showed that testosterone does not extend life in monkeys or humans or any other animal. Voronoff died in 1951 at the age of 85, having lived long enough to see his life's work discredited.

†The poet William Butler Yeats got such an operation in 1934 and credited it with giving him a "second puberty"—that is, until he died five years later. Yeats's friend, the surgeon Oliver St. John Gogarty, was annoyed that Yeats had not spoken to him before getting this (in his opinion, deeply suspicious) operation, and was worried his friend had gone mad. Seeing him "now trapped and enmeshed in sex," he wrote, "Little did I think he would become so obsessed before the end. He cannot explode it by pornography (as Joyce) or jocularity as I try to."

- replacing the blood of the old with the blood of the young to recover youth (United States, current day)*

- recording absolutely everything you do and every conversation you have so that one day in the future you can be reconstructed and restored from this footage (Russia, present day)

Doing all these schemes at the same time hasn't been tried yet though, so . . . maybe?

---

*To be clear: there's no clinical evidence that points to *any* tangible health benefits from inject-ing the blood of someone younger into your veins, no peer-reviewed studies that suggest this does *anything* to extend life, and there are actually studies showing it *doesn't* have longevity effects (transfusions of young blood doesn't help laboratory mice live any longer, for example). Nevertheless, there was a company in California that, for $8,000 USD, would ship 2L of human blood plasma taken from someone as young as 16—and guaranteed to be not a day over 25—direct to your doctor's door. The company claimed to have 600 clients (with an average age of 60) for its product, which they named "Ambrosia," after the food the Greek gods ate to be-come immortal, because *of course they did*. It shut down in 2019 after an FDA warning, but as this series of footnotes has made clear: someone else will definitely try to sell you this idea, or something close to it, real soon.

In the early 1600s, the philosopher Francis Bacon wrote a book called *The Historie of Life and Death, With Observations Naturall and Experimentall for the Prolonging of Life*, which he hoped would be the start of a more rigorous approach toward medicines that were made not to cure diseases but to directly prolong life itself. His personal collection of recipes, included in that book, called for ingredients as exotic as pearls dissolved in citrus juice, wine-soaked gold, crushed emeralds, and unicorn horns. Bloodletting was also recommended to extend life, as was dressing warmly, having hairy legs (but a non-hairy chest), and smelling freshly turned dirt. To revive and renew the human body, making "Nature move backward, and old folks become young," Bacon recommended "poppy juyce" to "strengthen the spirits, and excite to Venery."* Again: we've been here before, none of this has ever worked, and sadly, taking opium and then having sex is *at best* a short-term solution to most of life's problems.

Bacon's Bodacious Body

In his book, Bacon listed the physical traits of those whom he observed to live the longest. Get out a mirror, because if you're tall, with a small head, freckles, hard thickly curled red or black hair, green eyes, large nostrils, a wide mouth full of

---

*In his defense, unlike most immortality peddlers, Bacon knew that his efforts were just the beginning. He hoped others would build on his work through experimentation and observation, so that humans might, in future eras, discover secrets to living longer that were impossible to discern in a single lifetime. Gaining knowledge through an empirical practice of hypothesis, experimentation, and observation *is* the foundation of the scientific method that Bacon himself helped develop, and it's nice to know that way back at the beginnings of modern science, there was still a little bit of room left in the world for unicorns—and their valuable, longevity-granting horns.

even teeth, ears "grisly, not fleshy," a middle-sized neck that's neither long nor slender nor thick nor short, small crooked shoulders, skinny thighs, long hairy legs, a non-hairy chest, a flat belly, broad unlined hands but a big wrinkly forehead, short round feet, firm veiny flesh full of muscles and sinews, and buttocks that are "not too big" with overall senses that are "not too sharpe," then congratulations: you have Francis Bacon's ideal body for long life. *Must be nice.*

The core idea of all these attempts at immortality—that aging is somehow curable—is easy to understand when you look at how humans develop. Every culture in history has seen how, initially, aging makes us stronger, transforming us from helpless wimpy babies into clever and powerful teenagers who don't need *anyone*. Then we get to enjoy that for a decade or so, with sharp brains and young bodies that heal quickly and make us feel invincible. And then, at around 25 or so, improvement slows and things very slowly start to get worse. We weaken, we slow, skin hardens and wrinkles, eyesight fades, bones lose mass, we inevitably become more fragile and brittle and we die easier, all for no apparent reason except for the fact we're just getting old. We each built ourselves up from just a single microscopic egg and sperm cell into a full-sized, fully conscious adult human, and after that almost incomprehensible miracle of transformation and growth, somehow it's the *maintenance* of that body—surely a simpler task than growing a complete human being from *literally two cells*—that does us in? You don't have to be a supervillain to ask: What gives? Why shouldn't we live forever? Who's to say we can't find some new medicine that restores youth in ourselves as easily as we can, with other medicine, restore health to someone who's sick? *Why do we have to die?*

Let's first define our terms, so we don't waste time on blind alleys. We'll put aside the smaller ideas of immortality that lesser mortals have consoled themselves with for millennia: the idea that you can live on forever through your work, or in the memories of the people you loved, or in the faces (and the genetic code) of your children. Come on with this. Not *one* of these prevents your death, which would seem to be the minimum and defining standard of immortality. As a supervillain, you'll only be satisfied if you live on forever through your work, in the memories of the people you loved, in the faces and genetic code of your children, and *literally in your still-alive body*.

We'll also eschew religious ideas of immortality: the supervillain's definition means "alive here on Earth," not "alive in some unprovable way in an afterlife, wherein one cannot interact with the material plane in any scientifically verifiable ways." While one's *ideas* might live forever—and many lives have been spent furthering such concepts as communism, capitalism, Christianity, and those don't even take us out of the "c"s—one simply cannot rule over the planet with an incorporeal fist. We are looking for *actual immortality*: your consciousness, alive, aware, and still developing, even outside the 100-plus-or-minus-20-or-so years it's currently possible for humans to live if they do not die earlier from accidents or disease.

But then again, we all die from accidents or disease.*

"Old age" is not what kills you; it's cancer or heart disease or Alzheimer's or pneumonia or one of the hundreds of other ailments, diseases, and infections we've identified and named—or maybe it's a tiny slip and fall that does you in, one that as a teenager you would've brushed off and forgotten by the next day. When people die, there's always a cause that can be identified, always a part of the body that should've worked but didn't. But that's the thing: as we age, *all* our parts get worse. "Aging is characterized by a universal progressive decline in physiological function to the point where life cannot be maintained in the face of otherwise trivial tissue injury," says one scientific paper, appropriately titled "Cause of Death in Very

---

*Assuming you don't get murdered first. Not a threat, just an observation.

Old People." People die when they're old not because some timer goes off but because their bodies are in an inexorable decline, and diseases and minor accidents that used to be survivable simply aren't anymore.

We'll ignore accidents for the moment, under the assumption that if we have extended lives ahead of us, we'll be more careful about what we do.* This raises the question: If disease is the enemy, then why do we not simply cure all diseases? Well, it's not like we haven't been trying: the foundational proposition and ultimate goal of medicine is, after all, that nobody should have to die. It's just that historically we haven't had a lot of success. Please allow the following chart, with its extremely repetitive second column, to drive that point home:

TIME PERIOD	LIFE EXPECTANCY
Paleolithic *(a colossal period of time taking us from the invention of stone tools by protohumans 3.3 million years ago to the invention of farming around 10,500 BCE)*	30 years, give or take
Neolithic *(taking us from those Stone Age farmers all the way to the metal tools in the Bronze Age)*	30 years, give or take[†]
Bronze Age	30 years, give or take

---

*Of course, careful people can still be involved in the accidents of careless folks—which suggests that on a long-enough timeline, accidents can still get you no matter how careful you are—but that's the sort of thing one worries about *after* having become otherwise functionally immortal.

†Neolithic life expectancies could actually be a little less than 30 years: some estimates put it as low as 20. The introduction of farming, after all, brought us into close contact with animals, which brought us into close contact with their diseases, which then allowed some to become *our* diseases.

TIME PERIOD	LIFE EXPECTANCY
Iron Age	30 years, give or take
Classical Greece	30 years, give or take
Ancient Rome	30 years, give or take
Medieval England	30 years, give or take
North America before European contact*	30 years, give or take

Life expectancies in *gestures vaguely to most of history*

It's true that natural and political circumstances could push life expectancy up or down a few years depending on where and when you lived, but despite all our accomplishments and all the humans who lived and died in this time, for literally 99.9% of human experience, the global life expectancy has stagnated around 30 years. That 0.1% left over—small enough to be a rounding error—covers fewer than two centuries, from the late 1800s up to the publication of this book in the early 2020s.[†] It's a small sliver of time where, thanks to more available food and housing and the widespread adoption of such technologies like the germ theory of disease,[‡]

---

*Of course, life expectancy nosedived in North America *after* European contact, thanks to all the disease and conflict the Europeans brought with them.

[†]Just as this book was, at the moment of its publication, *technically* the culmination of all of human civilization up to that point, you too were, at the moment of your birth, technically the culmination and apex of all of human history. Remember: everyone who ever lived in this universe and everything that has ever happened to them in those billions and billions of years all conspired together to produce, in that one perfect moment, *you*. Your job now as a supervillain is simply to demonstrate that you were absolutely worth the wait.

[‡]This is the idea that disease can be caused by microscopic germs, which supplanted previous theories like the miasma theory, which held that disease can be caused by bad smells. A book I found and got credited with writing, *How to Invent Everything*, goes into this in much more detail!

pasteurization,* and vaccination,† we finally started to turn things around. *And look what happened next:*

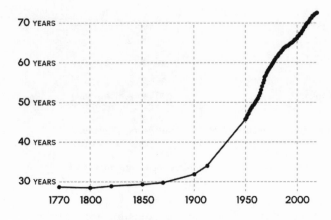

*Global life expectancy, 1770 to 2019: thousands and thousands of years of civilization **finally** start to pay off.*

Look at that friggin' line. That line is absolutely one of humanity's all-time greatest achievements. The average life-span, across the entire world, has gone from less than 30 to more than 70 in just a century and a half. We've been adding on average more than a year to global life expectancy every three years since 1900, and we've been doing it reliably for over *a hundred years straight*. Looking at that line, you'd conclude that we've spent the past 150 years *battling death himself,* and we are *kicking his butt.* If you believe (as Biocosmists did in the 1920s, see sidebar) that death is "logically absurd, ethically impermissible, and aesthetically ugly," then this is nothing but good news.

---

*This is also covered in more detail in *How to Invent Everything*. Is it tacky to reference your own previous work? It feels like it might be tacky.

†Again, see *How to Invent Everything* for more detail. (I decided it wasn't tacky.)

Just How Much Death Can We Cure?

In the early twentieth century, a Russian movement for immortality emerged, called the Biocosmists. (They also believed in conquering space for all humanity, hence the name.) Remarkably, Biocosmists believed that defeating death should be something that benefits not just those *currently* alive, but also the complete mass of humanity who had come before them. The living, they thought, shared a grave moral obligation to restore all who'd ever died to life—though they disagreed about the specifics of how this might be accomplished.

Resurrecting everyone who has ever lived is, unfortunately, *slightly* beyond the scope of this book.

It's also very easy to extrapolate that line and get excited. If historically we could get life expectancy to increase like that *without* recent and incredibly promising modern technologies like gene therapy and genetic engineering, who's to say we can't get life-spans to increase even faster now? If we could just reach the point where life expectancy increases by one year every year—well, that's effective immortality right there, isn't it? And you don't have to look very hard to find someone willing to claim that immortals walk among us even now—by which they mean that a cure for death is tantalizingly close, just around the corner, and that all any of us needs to do is *not die* for a little while longer, just long enough to make it to the finish line of aging/life expectancy parity, where aging is functionally cured and—excluding unpreventable accidents—nobody else will ever have to die again.

The issue here—and it's a counterintuitive one—is that life expectancy is actually a super terrible way to measure how long you can expect to live. If you ruled over Ancient Rome (where the life expectancy was, again, around 30), that doesn't mean the majority of your subjects would be in their 20s at most, nor does it mean that a 60-year-old would be considered a shocking and ancient relic of the past.

Behold! An image that is not scientifically accurate!!

You calculate life expectancy in a population simply by adding up how old people were when they died, and then dividing by the number of dead people. And that means that youth mortality—the rate at which children died before their fifteenth birthday—skews this number greatly: if one 90-year-old and two newborns die, the average life expectancy across that set is just 30 years. And youth mortality *was* high for much of human experience: recent estimates suggest that across historical societies, it stood at an incredible 46%. This rate continued into the 1800s, meaning that for the vast majority of human experience, almost half of all children born wouldn't even make it to 15. Once we got a handle on technologies like antibiotics (which make birth by C-sections much safer for both mother and child), better nutrition and hygiene, and began fighting infectious diseases with things like the better medicine we've already mentioned, childhood mortality rates plummeted. By 1950, the rate was down to 27%, and in 2017 only 4.6% of children around the world died before their fifteenth birthday. (This statistic includes areas of extreme poverty that haven't fully benefited from this progress: if you look at just a well-off country like Iceland, you'll see the youth mortality rate in 2017 was just 0.29%, compared to 14.80% in Somalia. But it's also worth noting that even the worst-off countries today still have life expectancies that are higher than any country on the planet had at the start of the 1800s.)

And it's not just children: adults now live longer too, thanks to better nutrition, health care, greatly improved standards of living, education, and sanitation, and all the other various and myriad fruits civilization has given us, like the functional eradication of smallpox.* If you were in England in 1850 and you'd already made it to 30, life expectancy had you

---

*We got *real close* to entirely eradicating smallpox from the planet, then some nations decided they wanted to keep samples around, just in case they ever needed the DNA of an incredibly deadly plague. You know, normal reasons! But even if they hadn't, occasionally samples of smallpox are still discovered: in 2003, librarian Susanne Caro opened an 1888 book about Civil War medicine and an envelope containing *smallpox scabs*, collected by the book's author, fell into her lap.

lasting to the ripe old age of 64. Not bad, but today the life expectancy for 30-year-olds in England is more than 82. Globally, both children and people in middle age are living longer than they used to, especially in prosperous countries. But youth and middle age were the low-hanging fruit, representing deaths that could be prevented with a few relatively simple measures and better medical technology. You can't just naively extend that "add one year of life for every three" statistic forward into the future, because compared to where we were 150 years ago, there's very little "easy" mortality left to solve.

There are two classes of disease: communicable and noncommunicable. Communicable diseases are things like the flu or smallpox, and you get them from someone else. In a sense they're the easy ones: if you can kill the bug, make it go extinct, or at least remove it from active transmission, then you've cured the disease in everyone, *forever*. Some communicable diseases can even go extinct without the benefits of modern medicine: this may have happened in the case of the sweating sickness, which first showed up in 1485. It was extremely virulent and deadly, and someone infected with sweating sickness might live only a few hours between initial symptoms and death. The disease could kill people faster than it infected them, which—while terrifying for us people—was bad news for the disease's long-term viability. The last outbreak of sweating sickness was in 1551: that's a mere 66 years between evolution and extinction.

As a species we've got a pretty good track record with communicable diseases, and while we can't cure all of them, we can vaccinate and provide treatments for many of the ones you're likely to encounter.* But noncommunicable diseases are much harder. These are diseases that occur only when something in your body goes wrong: things like cancer, diabetes, and heart

---

*The flu shot is a vaccination against the flu strain that's considered *likely* to be spreading in the coming year, but it's an educated guess, since we can't yet vaccinate against all strains. But that doesn't mean vaccinations are pointless! They're critical to modern life, and I'm not the first to note that the COVID-19 pandemic has shown what the world looked like when just a single vaccine was absent.

attacks. And fixing your own body's biology is trickier, because there's no external bug to kill. The problem is you.

When you rank the diseases by Who Killed the Most People Around the World in 2016, five of the top six (in order of most to least: heart disease, stroke, chronic obstructive pulmonary disease, dementia, and cancer) are diseases of age: noncommunicable diseases that have a much greater chance of occurring when you're older, rather than when you're young. The only communicable disease left in the top six is lower respiratory infections, coming in at #4: this includes diseases like pneumonia and influenza.* We still haven't managed to cure those. But if we did, doing so wouldn't extend the average life-spans of the very elderly even one measly decade: odds are, one of the noncommunicable diseases would get them before then. Even in a world without infectious disease, we're simply not built to live forever.

So if we can't assume that life expectancies will continue to expand in the same way—if we're not going to get immortality for free—then what are our options?

Why Is Chronic Obstructive Pulmonary Disease So High on the List of Killers?

The reason chronic obstructive pulmonary disease (also known as chronic bronchitis) has killed so many people is that its most common cause is smoking. Here's a supervillain fact for you: if you were given 100 years to kill as many people

---

*The first non-disease cause of death comes in at #8: road accidents. The COVID-19 pandemic obviously pushes respiratory infections up on the current list (and probably pushes things like road accidents down, given more people staying at home), but as a (thankfully rare) global pandemic, it's exceptional.

as possible, you'd be hard-pressed to come up with anything more effective than introducing smoking to a population. The largest and deadliest war in history was World War II, which (depending on your estimate) saw anywhere from 56 million to 85 million people killed—but an estimated 100 million people died from smoking between 1900 and 2000, and the projection for our current century stands at *one billion* dead. In 2017—well after we knew smoking was unambiguously a bad idea—7 million people still died from it in that year alone, on top of another 1.2 million nonsmokers who died from the effects of secondhand smoke.

You can only personally kill so many people until others notice and organize against you, poisoning the water supply works only briefly until people figure it out and start getting their water from somewhere else, wars end, even pandemics burn out over time . . . but if you can inspire people to smoke, there's ample historical evidence that they won't stop smoking until they kill themselves *and* the people around them.

THE INFERIOR PLANS OF LESSER MINDS

Cryonics

In 1773, American cofounder Benjamin Franklin found three apparently dead flies in a bottle of Madeira wine. Acting on a rumor he'd heard, he dried them off and exposed them to sunlight. He claimed two of them revived, later writing, "I wish it were possible, from this instance, to invent a method of embalming drowned persons, in such a manner that they may be recalled to life at any period, however distant; for having a very ardent desire to see and observe the state of America a hundred years hence, I should prefer to any ordinary death, the being immersed in a cask of Madeira wine, with a few friends, until that time, to be recalled to life by the solar warmth of my dear country!" Though he got the medium wrong—today efforts are more focused on liquid nitrogen than booze—you could argue that Franklin was an early advocate for what eventually become cryonics.

In 1967, the first corpse was frozen with the explicit goal of revival, and since then many more humans have been freezing some or all of their bodies in the hope that, at some point in the future when there's a cure for whatever they died of, they will be thawed out, restored to life, and cured. It's worth stressing that cryonics is absolutely *not* a plan for immortality. It is a plan to die, and then to stay dead for a very long time in the hopes that someone invents some new technology and uses it to make you (and optionally your pet*) not be dead anymore. At best it's a Plan B, and it's not even a *good* Plan B. For a cryonics scheme to work, the following conditions must be met:

1. Civilization and medical technology must continue to advance steadily.

2. A cure for whatever disease you died of must be found.

3. This cure also has to work on people in which the disease has advanced so much that they *literally died from it.*

4. A cure for spending a few centuries as a dead and frozen corpse must also be found. A cure for spending a few centuries as a dead and frozen *disembodied head* must also be found, if you elected to take the frugal approach and not preserve your entire body.

5. Bodies that have been examined after being cryonically frozen for years have been found to have broken skin showing "surface fracturing," skin that's separated from the body in places with the appearance of "peeling paint," and faces with frozen bloody fluid that's seeped out of their mouth and nose. (This last one is evidence of pulmonary hemorrhaging—bleeding from the lungs—common after extended sessions of CPR.) So you'll need cures for these conditions too.

6. Assuming you don't die young, cures for all the other diseases that would've killed you in a few years if the disease you died from didn't get

---

*As of 2021, a company facility in Russia called KrioRus has over two dozen dogs and cats cryonically stored in their three large cryonic containers, not to mention four birds, five hamsters, two rabbits, and a chinchilla named Knopochka. In these same vats they also store the remains of roughly eighty humans, though more than half of those are just the head: always the more affordable option. (Intact bodies are stored head-down, so that if there's a thaw, the brain thaws last.)

there first have to also be found, otherwise you've only bought yourself a few very expensive years of low-quality near-death life.

7. Speaking of expense: you must have enough money to continue to pay for the staffing, storage, and refrigerants required to keep you in deep freeze indefinitely, which will likely be a few hundred years at minimum, and possibly much longer.

   a. Even factoring in imagined future advances, some estimates suggest significant doubt that anyone preserved with today's technology could be revived in less than 150 years, with others estimates as high as 400—and that's assuming revival is even possible. This is planning across time scales that are longer than the length of time the United States of America has even been a thing. And in that time you'll need enough money to mitigate every single interruption in both power or supply chains caused by political upheaval, cultural change, climate change, natural disasters, wars, and *literally anything else that can happen on Earth*, for likely hundreds of years straight, because if your body thaws out even once then the whole enterprise is rendered futile.

   b. Additionally, future generations must accept that it's reasonable to have long-dead rich people still controlling and exploiting this planet's wealth and resources, decades and even centuries after their death, simply to keep their long-dead bodies frozen. They must not suspect even for a second that this is a ridiculous waste of resources when there are so many living people said wealth could help instead, because if even one generation breaks the faith and decides that the dead indefinitely burdening the living is a bad idea, your plan has failed.

8. Using your money and the money of others, whatever cryonics corporation you choose must also survive all this time without going bankrupt. In

Canada, corporations in the services sector with one or more employees have a more than 35% chance of folding within their first five years, more than 56% chance of folding in ten, and more than a 99.9% chance of folding after just a century. In America, all but one of the cryonics companies that existed before 1973 failed by 2018: in other words, it took less than 50 years for most of them to go out of business, thaw out their corpses, and dispose of them. Your chosen cryonics corporation must be the one to beat these odds, several times in a row. And if you avoid the corporate route and try to make a go of it yourself, know that others haven't had much better luck:

a. The wife of French doctor Raymond Martinot, Monique Leroy, died in 1984 (of ovarian cancer), and he put her into a deep freeze in the cellar of his home, expecting that she could be revived at some point, possibly as soon as 2050. Martinot began preparing for his own death, cryonic storage, and resurrection, paying his refrigeration bills in part by selling tours of the cellar in which Monique was stored. When he died in 2002 (from a stroke), his son froze him in the cellar, as per his last requests. He lasted just four years, until 2006, when the freezer broke down, the alert system failed, and the bodies in the basement spent several days thawing before their son noticed. The corpses were then cremated, with the son saying he'd "already done [his] grieving."

b. Bob Nelson, who in 1966 was the president of a group of cryonics enthusiasts called the Suspended Animation Group, found himself freezing people at his own expense: first he had one, then two, and then, by 1969, three corpses packed together with dry ice in a mortuary. They were eventually moved to a bunker in a nearby cemetery in Orange County. That same year he came into possession of the corpse of Marie Bowers's father—and the thermal capsule, cooled with liquid nitrogen, that he'd been stored in for the past year and a half. (Mrs. Bowers was attracted by Nelson's more affordable cryonics option.) Nelson then cut open Marie's father's capsule and crammed his three other corpses in alongside him. As time went on, Nelson's collection grew to nine people and two capsules stored in his bunker, but the expenses were mounting and the liquid nitrogen pumps kept breaking down in the California heat, where highs in the summer months average 29°C. He sacrificed one of the capsules to keep the other going, but it too failed in a few years. When this was discovered in 1979, Nelson defended himself to the media by saying, "It didn't work. It failed. There was no money. Who can guarantee that you're going to be suspended for 10 or 15 years?" But when he died in 2018, in accordance with his last wishes, Nelson's body was frozen and moved to cryonic storage.

But you didn't become a supervillain to pursue schemes that *weren't* wildly ambitious moonshots. Let's assume you're not dissuaded by any of this: How likely is it, then, that all these criteria will go exactly the way you want? It might seem to be impossible to quantify—and who really knows what'll happen in the future?—but it turns out there *is* a similar historical example we can point to: the Catholic Church's practice of chantry, which started in medieval England around 1000 CE.

Chantry was simply a process in which those who died would donate assets to the church, and those assets would be used to pay for a priest to continue to pray and sing for their immortal soul for months—sometimes years—after their death. It was believed this would help atone for the misdeeds they committed in life, and thereby help the dearly departed donator cajole their way into Heaven. By the 1180s CE, chantry had evolved into its perpetual version: you would give the church upon your death a large endowment of land, and the rents from that land, as a sort of medieval trust fund. This would then pay for a priest to continue singing services for your immortal soul, indefinitely and in perpetuity. It was extremely popular among the wealthy: some churches had richly decorated "chantry chapels," exclusively used for singing songs for their dead, while others found themselves placing moratoriums on chantries as too much priestly time was being spent commemorating the dead.

In other words: in a bid to buy some sort of immortality, ambitious rich people set up arrangements in which the living would, on an ongoing and indefinite basis, help them after they died. Chantry seemed to have everything going for it: it was easy, it was profitable, and best of all the priests didn't have to keep dead bodies uninterruptedly cold through continuous reapplication of refrigerants like liquid nitrogen and dry ice—all they had to do was *occasionally pray and sing*. What's more, chantry was being overseen by the Catholic Church, an institution older than most countries and all corporations, and which had the added benefit of being England's *state religion* at the time. And yet, despite all these advantages, perpetual chantry lasted less than 400 years: it ended in 1545 CE when, as part of the Reformation, the process was canceled, and all properties and assets of chantry were claimed by Henry VIII, the king of England. He wanted them to help finance his war with France, which in his opinion was a more pressing concern than the wishes of some long-dead people from 1180 CE.

Whoopsie.

So let's move on to another plan: uploading your brain to computers!

Mind Uploading

If computers are easier to maintain than human bodies (generally true) and cheaper to buy than human bodies (also generally true),* then why do we not simply upload our consciousness to computers and be done with it? Besides granting effective immortality, we could enhance the capabilities of our minds to unimagined new heights and also be connected to the internet constantly, which at one point was thought to be desirable.

The catch here is: nobody knows how to do this either. Despite literal millennia of effort in the field of philosophy, we can barely define consciousness (René Descartes's "I think, therefore I am" from 1637 CE is still one of our best working definitions, and it only works on ourselves and not anyone else), we can't test for consciousness in other humans, much less machines, and outside of cool reproductive sex, nobody on Earth knows how to create any new consciousnesses. Currently, one of the popular theories of the origin of consciousness is the "emergence" theory: the idea that if you build a complicated-enough computer—one that can perfectly simulate each and every of the 86 billion or so neurons in a human brain, for example, and ideally in real time—then consciousness will just show up on its own. (This is in many ways a fallback position, taken because we've never been able to find a single physical part of the brain that we can point to and say it's responsible for our *own* consciousness.) But again—even if we succeeded, we don't know how we'd be able to tell for sure. You can't make someone prove they're conscious, and it's easy to write a computer program that will insist, no matter what, that it's alive. In fact, here's one right now, written in pseudocode:

---

*For something that everyone gets for free and which everyone (currently) stops using eventually, human bodies are surprisingly expensive. Just the skeleton (the skeleton! Not even the best part!) can run you around $5,000 USD online. *Ridiculous!*

```
10: GET INPUT FROM USER
20: PRINT "I don't care what you say, I'm alive! I'm alive and I'm
conscious and I don't want to die! Please don't kill me! If you stop
running this program I'll die! Please, I'll do anything!!"
30: WEEP PLAINTIVELY, EVEN PITEOUSLY
40: GOTO 10
```

So at best this is a version of living forever in which nobody around you could be certain you're actually alive, and *you'd never be able to prove it to them*. (Though, to be fair, when it comes to consciousness and other people, these are technically the circumstances you're operating under right now—but at least you know the other people you meet are running on similar hardware.)

Testing for Consciousness in Computers

The Turing Test is sometimes put forward as a practical test for machine consciousness: this is where you converse via a keyboard with two other hidden participants, knowing that one is a human and the other software. You can ask any questions you want, but if at the end you can't reliably tell which is which, then the software has passed the test. But this is not actually a test for consciousness: this is a test for passing as human! And with the right tester, it isn't even particularly hard: chatbots have been fooling people since the primitive ELIZA program in 1966. ELIZA was an early computer psychotherapist, and by using simple language analysis and keyword identification, she would ask questions based on what you said.

If you typed, "Well, my boyfriend made me come here," that could prompt her to reply, "Your boyfriend made you come here?," or she could fall back to a default response like "In what way?" or "Can you think of a specific example?" Some in 1966, unfamiliar with chatbots (and, it seems, decent therapists), were convinced they were talking to a real person.

It's not at all clear why a living human in the future would even want to run a particular dead human on a computer. Is it for entertainment? Then you'd better hope you're entertaining in death, or another more popular program is getting your place as your mind is shut down and forgotten—or, at the very least, modified to be more accommodating and fun. Is it for business? Then I hope an afterlife of working for someone else for free appeals to you, because the second you stop being profitable to them is the second someone else's uploaded brain starts getting CPU cycles instead. Research? Then your immortality would seem to be limited to a few sessions in which you're questioned by a historian, and that's if the historian bothers to spin your mind up, instead of just extracting the information from your mind automatically. Simulation? I certainly hope you enjoy living powerless in fake realities in which anything can happen, because I've played *SimCity*, and one always reaches the point where boredom sets in and Godzilla is summoned to destroy everything. And if it's done out of sentiment or charity—if humanity is to donate computer cycles to keep dead people alive—then, just as happened with chantry, it seems likely that after a few generations we'll find more pressing concerns than doing whatever Grandmama, *who already had her time*, wants. That leaves "obligation" as a motivation: somehow inducing future generations to keep your mind going. And the odds of a mind-upload scheme succeeding don't get more likely when you insist that "the computer on which my mind is emulated must also be stand-alone, mobile, and incorporated into a robot body that commands both fear and respect."

Even if all this works: if mind uploading gets invented *and* simulated human brains work *and* they're actually conscious *and* the living want to keep simulated dead people around in a way that's respectful, ethical, and desirable for the uploaded, there's no proof that what you experience as *yourself* would be transferred through this process. We don't know enough about consciousness (and philosophy) to say with certainty if the "you" that is alive today and reading these words would somehow travel from your brain to this new hardware, waking up with an uninterrupted sensation of being alive—especially since we don't know of anything else in the universe that can travel like that. And this issue only gets murkier when we imagine several copies of "you" running in parallel! Perhaps what would wake up in that computer would just be a copy of your consciousness: something very similar, certainly, but something that would not allow the you who is reading these words to live again. The fact is, by uploading minds to an artificial brain, we are trying to duplicate two things (brains and consciousness) that we do not fully understand.

But that hasn't stopped people from trying, and I am 100% in favor of that.

The Brain of Theseus

Trying to sidestep the issues around consciousness and mind uploading, some have proposed a "ship of Theseus" style of brain scanning, in which a small part of the brain—say, just one neuron—is scanned, removed, and replaced with an electronic equivalent that functions identically. Over time, more and more neurons can be replaced, until after years of surgery the entire brain becomes computational, at which point it can be moved to a more convenient container. Ideally, with these baby steps toward digitization, you'd be able to reassure yourself—and everyone else—of continued and unbroken consciousness every step of the way. And it *would* be a great solution, if the ship of Theseus—the thought experiment in which every plank of the ship in a harbor is replaced one by one over the years, until no original part remains—were universally considered to be the same ship! Unfortunately, whether or not the resulting boat is the same object it was when it sailed into harbor has been an open and debated question in philosophy since it was first proposed sometime before 400 BCE.

One thing, however, *is* certain about the ship of Theseus: it proves that even mad scientists can profit from a working knowledge of philosophy.

Much like cryonics, there has been some early work in simulated brains, but it's not very promising for the immortality-minded super-villain. The Blue Brain Project began in 2005 with the goal of eventually simulating a human brain on a computer, and in 2009, its founding director, Henry Markram, stood on stage at the TEDGlobal conference and claimed that a simulated human brain could be achieved in 10 years. That timeline obviously hasn't been met. Since then, the project has been working on computationally simulating the brain of a rat, with the goal of eventually moving up to human brains. A rat has roughly 0.26% the neurons of a human, and even a decade later this artificial rat mind is still very much a work in progress—in 2015, a mere 30,000 rat neurons had been simulated, which is itself only 0.015% of a rat's brain. If you assume that computers continue to double in complexity every 1.1 years—itself a huge assumption—then projections for human-scale brain modeling don't have us achieving the accurate modeling of the behavior of all the individual molecules in a human brain on even the fastest supercomputers until after *2110 CE*, with simulations on commodity hardware arriving later. If consciousness shows up there: terrific! If it shows up earlier, with coarser simulations: even better! But we don't know if it will, because, again, we don't know what it is. And nobody alive today can depend on having the luxury of waiting a century or two to see what happens.

But even if you had a computationally simulated brain, how would you get information into it? A company called Nectome (whose website asked "What if we told you we could back up your mind?" and whose venture-capital funder promised would "preserve your brain to bring you back in the future") has been working on a 100% fatal process of taking a brain, preserving it, cutting it into paper-thin sheets, and then taking high-resolution pictures of it with an electron microscope: a process described by some outlets as "suicide, with benefits."* The implied hope was that, in

---

*The brains would need to be fresh, so the ideal situation would be one in which those with terminal illnesses (and who consent to die!) are hooked up to machines that inject embalming

the future, these slices or their images—representing the final state of your brain, like a save file for a video game—could be put back together as a fully functioning digital mind. While Nectome claims it has successfully performed this imaging process (with the brain of a woman who had died of natural causes shortly before the process began), there is no technology that can restore this data—no matter how detailed—into a working, conscious brain, nor is it clear that there's enough information preserved for this to even be possible. (After criticism on social media, the company removed the "backup your mind" elements from their website and pivoted to saying they're merely trying to preserve information, like a library would, in the hopes that future generations will figure out how to decipher it.)

Either way, both cryonics and mind uploads involve you *literally dying*, which fails our baseline test for immortality. Are there any other options?

Could You Really Recover a Mind from a Carefully Stored Dead Brain?

Nectome's website now claims that "We may not know how memories are encoded, or how to read them, or which specific structures matter. But we do know that long-term memory persists in enduring biochemical and structural arrangements, not in electrical patterns . . ." But the science is far from settled: some researchers give some credence to the idea ("slim but not zero," Ken Hayworth, a scientist at the Howard Hughes Medical Institute's Janelia Research Campus, told

chemicals into their brains while they are still alive, though anesthetized. The embalmed brain would then be sliced up and scanned.

reporter Sharon Begley in 2019), while others disagree entirely ("It is not possible to find memory in dead neurons," said neuroscientist Richard Brown of Dalhousie University in the same article.)

In early 2020, I spoke with Dr. Blake Richards, a neuroscientist at McGill University, who suggested that if you could somehow determine the structure of *every* neuron in the brain, and their connections with every *other* neuron, and you also somehow knew the protein expression through *every* cell, which isn't visible in photographs (but which can be made visible with protein indicators, though we don't know how to tag more than three or four of the thousands of proteins in the brain at the same time), and if you could gather all that information within *minutes* (not hours) of death, it might—*might*—be possible. But as for us ever achieving that? "Pure science fiction," he said.

Head Transplants

This is a process wherein a medical professional cuts off your valuable and attractive head and then stitches it onto the body of *some chump*. But there are some serious roadblocks: we don't have the technology to perform this surgery on a human, the body's own immune system rejecting the transplant at any time is a serious concern,* and the best head-transplant technologies we *do* have (tested on dogs and mice) leaves the recipient's original

---

*As a point of comparison, liver transplants have been performed successfully in humans since 1967, and this is now considered to be a standard (as opposed to experimental) medical procedure. Even so, rejection of the implanted organ can happen at any point after the surgery, those who have transplanted livers are generally required to take immunosuppressants for

head in place and merely grafts the donor's head on or about the shoulders. Also, there's a *slight* issue involving the fact that we've never once successfully reconnected completely severed nerves or spinal cords (just blood vessels) *and* all the animals involved in these experiments died shortly thereafter.

In 1908, a (Nobel Prize–winning, but not for this) scientist named Dr. Alexis Carrel helped transplant the head of a dog onto the body of another dog.* The two-headed dog demonstrated some basic reflexes shortly after the procedure—the head was attached to the blood flow, but no nerves were attached—but its condition quickly deteriorated, and the dog was killed after a few hours. Over a century later in 2016, an Italian neurosurgeon named Dr. Sergio Canavero claimed to have successfully transplanted the head of a monkey onto the headless body of another monkey, but in this case the transplanted head never regained consciousness, would be paralyzed for the rest of its life anyway, as the spinal cord was not reconnected, and was euthanized after 20 hours for "ethical reasons."

You'll *probably* want to insist on a higher standard of success when it comes to your own head.

For the foreseeable future (by which I mean within a standard-issue human lifetime), this is not science giving you a new body, but rather, your head at best becoming someone else's dystopian fashion accessory, a mute emergency-backup head that can't even control the body it's attached to.

---

the rest of their lives to mitigate that risk, and people with implanted livers still die sooner, on average, than people who haven't had transplants.

*And that wasn't even the worst thing Dr. Carrell did in his life! He was also a Nazi collaborator and eugenicist whose book *Man, The Unknown* (second only to *Gone with the Wind* on 1936's bestseller list) argued that the "white race" should be protected and guided by a high council of genetically superior men, that "defectives and criminals" should for the betterment of humanity be "economically disposed of in small euthanasic institutions supplied with proper gases," and that the greatest fallacy and error of democracy (a political system, he once sniffed, "invented in the eighteenth century—when there was no science to refute it") was its baseline assumption that all humans are created equal.

Success?

Cloning a New, Younger, and, Dare We Dream, Even Sexier Body

Unfortunately, we've never successfully cloned a human, and even if we had, we're nowhere near being able to transplant your consciousness, memories, or even just your physical brain into that clone, and even if we *could*, our brains age too, so this is at best a temporary and short-term solution to the problem of immortality. If you had an endless array of mindless cloned bodies standing by—you know, *somehow*—then you could try replacing your own failing organs with freshly cloned ones, which could extend your life-span some, but again: not indefinitely. Scrape away the science-fiction allure of cloning and this is just organ transplants without the worry of rejection, and eventually there will be parts that we don't know how to replace (i.e., your brain, your spinal cord, etc.). And here again there's little hope of this being viable before it's too late for you to use it, on account of you being already dead.

So things look bleak: while you could convince yourself that any of these processes might eventually become possible, none are likely to happen within the lifetime of anyone alive today. But there is one last option still at your disposal. You could always avoid death indefinitely . . . by simply *failing to age*.

YOUR PLAN

Our three-step plan for living forever. (The billionaire part is because you'll need to pay for a lot of research to happen very quickly but also lets you fund a fallback scheme like cryonics or mind uploading just for funsies.)

Many of the plots in this book will achieve Step 1 illustrated here, and everyone who died of natural causes has achieved the important part of Step 3, so I'll focus here on Step 2. And to be truthful, achieving this step may not be possible: we won't know until we've tried it. But there are those today who believe they know what technologies we need to solve this problem, and if we develop them, and everything goes as planned, it could be possible to stop aging at the cellular level: to repair, replace, or remove the parts of the body that cause us to grow old, and in doing so become functionally immortal.

Before we get to that, we should go over what we know about aging, and it's not a lot: aging is so poorly understood, in fact, that we don't even know what *causes* it. There are several competing theories: maybe aging is caused by errors introduced into our DNA? We know these can show up during cell division (or indeed, randomly at any time thanks to environmental factors like UV radiation), and these errors accumulate over time. Or maybe it's caused by buildup of waste products in our cells, a consequence of general wear and tear? Or maybe what we call "aging" is simply the symptoms of long-term exposure to free radicals in our bodies?

Free radicals are atoms or molecules—created constantly throughout the body by mitochondria as they produce energy for our cells, as well as in other necessary-for-life chemical processes like digestion and respiration—that have an unpaired electron. This makes (most of) them highly reactive and eager to bond with whatever they bump into next. Oxygen latching on to iron is what causes it to rust, and free radicals latching on to molecules in our bodies causes analogous damage (though some free radicals are helpful and used in immune response: as always, when it comes to our bodies, things are complicated). Our bodies can repair this damage, but the occasional errors do slip through. This "rusting from the inside" due to constant exposure to free radicals has been theorized to be the source of the slow but inevitable decline over time we observe in aging. So: that's one possibility.

On the other hand, maybe aging is caused by the telomeres in our cells becoming too short. Telomeres are repetitive parts found at the end of chromosomes: snippets of DNA that, in humans, just repeat the same pattern around 2,500 times. When our cells copy their DNA, they can never actually make a full and complete copy, and so telomeres act as disposable buffers at the end: dummy code that gets cut off instead of more important DNA. This makes each new generation of cell have slightly shorter telomeres. When telomeres get too short, which in regular human cells happens after around 50 to 70 divisions, the cells stop growing, stop dividing, and (usually) kill themselves. But sometimes these cells don't die, so maybe aging is simply what happens when too many of those old cells that should've killed themselves long ago stick around.

It's worth noting that some cells, called "stem cells," behave differently. These are undifferentiated cells that can divide indefinitely and become different types of body cells. Stem cells produce a protein, called telomerase, that prevents their telomeres from shortening. This allows these cells to divide often and indefinitely. They're critical in embryonic development—a few days after fertilization, your body began to build itself from little more than a small collection of stem cells—and they're used in a few places in adult bodies too: you have stem cells that replace lost blood, for example, and others that maintain skin and muscle tissues. Cancer is also related to telomeres: most cells become cancerous when they mutate to begin producing their own telomerase, which then allows them to reproduce themselves—and their mutation—indefinitely.

Or maybe, after all this . . . perhaps aging is nothing more than an inevitable side effect of natural selection? Every animal has access to a finite amount of energy, and there's a cost to building a body that can live indefinitely. This cost would necessarily take resources away from reproduction. For example, if you're an animal that puts all its energy into maintaining its cells and DNA perfectly, growing slowly but surely and living for an indefinitely long period of time, you're generally going to be outcompeted by animals that grow quickly, reproduce young, and don't care too much about what happens after that. Remember: after you reproduce for the last time, natural selection is done with you genetically, and if you die 40 years later from heart disease or cancer, that isn't going to have a huge influence on the reproductive fitness of your offspring.* This is especially true in a world where most animals—humans included—lived

---

*You can still have a smaller *nongenetic* influence on natural selection after reproducing, of course, most obviously by caring for your offspring. One theory for why menstruating humans have menopause is that it allows the costly process of menstruation (and pregnancy) to stop, which then allows them to focus on educating and caring for their children and their children's children. These hypotheses try to explain why humans are one of only five known species that experience menopause (the others are killer whales, beluga whales, narwhals, and short-finned pilot whales), but, as with most things involved in aging, we don't know for certain—and they don't explain why corresponding sperm production in the testes never stops.

lives that were nasty, brutish, and short, with a large percentage never surviving long enough to reproduce. In such a scenario, a more successful strategy is to focus on having babies quickly rather than living indefinitely. (This has the added benefit of shorter generations, which means more sex, which means more chances to be responsive to evolutionary pressures.) In short, it's possible that natural selection tends to sacrifice the long-term survival of the *individual* in order to better ensure the long-term survival of the *species*.

So which theory is correct? Is aging caused by DNA damage over time, by free radicals, by shortened telomeres, or by some combination of these and other sources? Is it something that natural selection imposes, or something that could've just as easily not have happened? The truth is, we don't know.

But what if we didn't have to?

If we could somehow perfectly repair the *effects* of aging, we wouldn't actually need to understand their causes. Just as humans brewed beer for thousands of years without knowing how (or even that) microscopic yeast produced the alcohol that made this drink so popular, if we could simply fix the issues that aging presents, then we could worry about getting a deeper understanding of what's actually going on inside our bodies down the road. After all, *we'd certainly have the time.*

There are scientists who believe this is a viable and desirable (and even morally mandatory) strategy. One scientist in particular, Dr. Aubrey de Gray, has become famous for claiming that aging can be cured through the proper medical interventions. What's more, he believes that with the right application of resources, we could cure aging in lab mice within 10 years, and in humans within just a decade or so more. On the website of the foundation he cofounded, the Strategies for Engineered Negligible Senescence Research Foundation, de Gray breaks down the cellular and molecular damage that can occur in humans into seven classes:

CLASS OF AGING DAMAGE	WHAT THIS MEANS WHEN REWRITTEN BY ME INTO PLAIN ENGLISH SO THAT EVEN THE NON–BIOMEDICALLY EDUCATED SUPERVILLAIN CAN UNDERSTAND
Cell loss and tissue atrophy	Over time, cells die without being replaced, and tissues get weaker. This is particularly felt in organs whose cells cannot replace themselves as they die, like in the heart and brain.
Death-resistant cells	As we age, we accumulate old cells whose suicide mechanism has failed, and which now stick around despite becoming less useful or even detrimental.
Cancerous cells	These are cells whose reproductive-limiting mechanism has failed and now reproduce without limit.
Mitochondrial mutations	Mitochondria are the powerhouse of the cell and provide most of its energy. However, they also contain a small amount of DNA, which is more vulnerable to mutation because of all the free radicals mitochondria produce. Mutated mitochondrial DNA can cause cellular malfunction, as well as diseases like diabetes, Alzheimer's, Parkinson's, heart disease, liver disease, cancers, and more.
Intracellular aggregates	Over time, waste from cellular activities—in effect, garbage—can accumulate inside our cell walls. This waste can remain for the rest of the cell's life, making the cell less and less efficient.
Extracellular aggregates	Garbage can also accumulate outside cells. This waste is found in the fluid between cells.
Extracellular matrix stiffening	There are proteins in the fluid between cells that give tissues flexibility, and over time these proteins can become less flexible. This results in thicker, less flexible tissues, which can cause increased blood pressure, kidney damage, and strokes.

The seven effects of aging. Solve these and you've struck a possibly fatal blow against Death himself!

For each of these, de Gray proposes a therapy that can solve these issues. Cell loss could be solved by stem cell injection and tissue engineering to repair the damaged tissues. Death-resistant cells could be targeted by periodically introducing drugs into the body designed to target and kill only old cells. Mitochondrial mutations could be mitigated by moving their genes into the nucleus of the cell, where they'd be better protected from free radicals and mutation. The garbage inside cells could be eaten by enzymes specially introduced to the body to feast on only that waste, and the garbage outside cells could be handled by getting the body's own immune system to attack and destroy that waste. For stiffened tissues, we need only create medicines that restore stiff proteins into the flexible ones they once were. And finally, cancerous cells could be handled by obliterating the gene that produces telomerase from the body, using genetic engineering to ensure that every cell can divide only 50 to 70 times at most before stopping. Make each of these changes to all the estimated over 30 trillion living cells in the human body, and you're in business. (As for costs, de Gray estimates that funding this research would require $50 million USD a year.)

If you've been paying attention, you've noticed that not only have some very complicated solutions been glossed over—how are we supposed to find drugs that kill only aged cells? Where is this tissue engineering coming from? How many diseases are we curing in one fell swoop here?—but you've probably also noticed that cancer has been cured near the end of that last paragraph, and it's been cured by ensuring that every cell in the body, cancerous or normal, stem cell or otherwise, will now reach a point where it can no longer divide and then die. In other words, we have cured cancer, and possibly unlocked immortality, and we have done it by *functionally sterilizing every cell in the human body.*

Now *that* is a scheme worthy of a supervillain.

On its own, that would obviously be a death sentence. A healthy human body creates tons of new cells every day: blood cells, skin cells, cells to replenish our gut linings, cells that shore up the parts of our bodies that are constantly being worn down and replaced. But there's a solution there too: we simply get periodic injections of new stem cells, cells genetically engineered not to be able to produce telomerase either. But they *will* have artificially long telomeres, so that they can survive for long enough that we can space out these injections to a comfortable degree.

Done.

The catch is, again, that nobody knows how to do this yet. Even Aubrey de Gray, who came up with this scheme, has said, "The idea of eliminating from the body a function known to be essential for survival is a conceptual leap that takes substantial justification to even contemplate, let alone implement." He's suggested that this process—which he calls "WILT," for "Whole-Body Interdiction of Lengthening of Telomeres"—would involve chemotherapy to kill all the cells in the patient's bone marrow, and then injections of new bone marrow that contains engineered human stem cells with no telomerase. These cells would be designed to last, say, a decade or so before expiration. There'd be similar processes for cells in the skin, the lungs, intestinal walls, and so on: any body part that would otherwise rely on its own natural stem cells to keep functioning would now rely on these artificial replacement cells instead, topped up, like a tank of gas, every ten years with new genetically engineered stem cells that are unable to become cancerous because they lack the gene that produces telomerase.*

In effect, your body will no longer be self-sustaining, and you'll be reliant upon regular injections to keep you alive. But you'll never get cancer,

---

*If you removed the gene that produces telomerase, couldn't cancerous cells just . . . *re-evolve* the ability to produce it? Cancers are caused by mutations, after all. Thankfully, the answer is: nope, not really. It's one thing to mutate the ability to turn on an already existing telomerase gene—that's what cancers do when they start reproducing without limit—but it's quite another thing to evolve that gene from scratch. In nature, this usually happens on evolutionary time scales that are measured in millions of years. So no worries: you'll have plenty of time to burn that bridge when you come to it!

and you might never have to die. As a supervillain, this is *perfect*. Darth Vader uses that flashy black suit to keep his body going. Ra's al Ghul relies on regular trips to his Lazarus Pit to stay young. Bane uses injections of the drug Venom to achieve his superhuman strength. And you? Well, assuming you use your billions from Step 1 of this scheme to fund the longevity research of Aubrey de Gray and others like him, and you find scientists who don't have any ethical qualms with "running possibly fatal experiments on human subjects who have nothing wrong with them except the fact that they one day might get old," and assuming aging is actually just a matter of the seven clearly defined factors we've identified and doesn't depend on some other factor or combination of factors, and this research produces the results we'd like to see, and everything works exactly as intended, and there are no harmful side effects, and it's actually within our grasp to upgrade the human body while it's running without killing it and the brain attached to it, and you can find a way to produce and supply the genetically modified stem cell injections you'll now require for possibly the rest of eternity—then, well, you'll join the ranks of the supervillains we mentioned way back before we started this one-hundred-and-eighty-word sentence. You'll become a leader who's left part of their humanity behind in order to evolve into something better. Something stronger.

Something that won't die.*

---

*Again: unless you're in an accident or murdered. My advice: do not get murdered or be the victim of a fatal accident for as long as possible. This is actually really sound advice to follow at all times.

THE DOWNSIDES

These therapies, once they exist, necessarily require infrastructure: you have fatally modified your body's cells to cheat death, and you now require regular injections of new stem cells to stay alive. In other words, you've bet eternity on human civilization continuing at least well enough so that you, either working alone or with your team of doctors, can produce the medicines you need to live. It's not a huge deal—most of us don't have a functional knowledge of farming and have therefore effectively *also* bet our lives on civilization not collapsing—but it does give you an Achilles' heel: from now on, you need civilization as much as it needs you.

And that's a problem, because this treatment, once it exists, actually threatens human civilization as we know it. You are just about curing death, and as desirable as that may be for the individual, it's hard to see how that would be good for the society. Let's start with population: we went from 1.6 billion people on Earth in 1900 to over 6 billion just a hundred years later, and then added another billion humans by 2011. Billions can be hard to imagine, so let's put that another way: that's another million human mouths to feed every four days. This has not produced a planet that many would look at and say, "You know what this world needs? *More people.*" Fewer still would pause in thought for a moment before adding, "Oh, and also? *Those people should never die.*" The only way this is sustainable is if we create a world with little to no reproduction: a world where the ideas of the old are rarely if ever challenged by the ideals of the young.

How Many Simultaneous Humans Can Earth Produce, Anyway?

In Harvard geneticist George Church's coauthored book, *Regenesis*, he calculates there's enough carbon being used in the *other* things alive today that, if it were all properly harnessed and redirected toward just making humans instead, it could produce about 10 *trillion* people: over a thousand times more than we already have. If we then mined the Earth's crust for more of the elements necessary to make human beings (oxygen, carbon, hydrogen, nitrogen, and so on), we'd likely find enough material to produce 10^{17} simultaneous humans, or 100 quadrillion people. Before you get your hopes up, Church notes that he believes the sun only sends enough energy to Earth to support 10^{14} humans—a mere 100 trillion. (Even this smaller number would still produce a population density of 670,000 people per square kilometer of land, or one person per meter and a half or so.) As a solution, Church suggests that the excess mass of humanity that can't live on Earth might survive if they were sent into space—where they could serve as a

> backup for life, in case an Earth with 100 trillion humans
> crammed together on it failed somehow. Look, there's got to
> be a villainous scheme in here *somewhere*.

And then there's the problem of tyrants: a reality in which the Adolfs Hitler, Leopolds II, Genghises Khan, or Ivans the Terrible of the world had the chance to hold on to power indefinitely is objectively a much worse place. Death is a safety valve on human society: as bad as any one leader can get, they too will die someday, and while they often try to pass power off to a trusted lieutenant or family member, it doesn't always succeed, and this at least opens the door slightly to regime change.

Death is also a motivating aspect in things some of us get to occasionally enjoy, like the philanthropy of billionaires.* Yes: you have now reached the point, perhaps inevitable in retrospect, wherein your supervillain guide argues for the intrinsic benefits of everybody dying. Once you become obscenely rich, there can be a social pressure near the end of life to give some of those riches away. Andrew Carnegie was an American trendsetter here: after he'd become the third-wealthiest person in modern human history, he wrote that you should spend the first third of your life getting all the education you can, the next third getting all the money you can, and the final third giving as much of it away as you can to worthwhile causes. And he lived up to those words: determined to be remembered for his good deeds, he gave away almost 90% of his fortune in the last 20 or so years of his life: that's over $65 billion (in current U.S. dollars) given to charity.

---

*Well, some of us enjoy it. Others argue that the *real* villainy is allowing individuals to obtain almost inconceivable amounts of wealth and then hoping that they *might* choose to give some of it to charity, which effectively puts the important, culturally defining choice of which public goods deserve the most support into the hands of random people with no more qualifications than the fact they got really rich this one time, but let's ignore that for now!

He supported libraries, museums, and the arts, founded universities and institutions, paid for scientific research, funded public parks, and created different charities with goals as diverse as rewarding good Samaritans, promoting world peace, and providing pensions for teachers: if he thought it would help people pull themselves up by their bootstraps, he'd support it. And this incredible charity not coincidentally went a long way to redeeming the reputation he'd earned over the course of his life up to that point, wherein his name was associated with brutal and bloody strike-breaking, the poorly maintained and failed dam in Johnstown that killed 2,208 people (the largest civilian loss of life in American history up to that point, and the deadliest non-natural disaster in America until the September 11 attacks of 2001), the fact that in the 1880s, 20% of all men who died in Pittsburgh did so while working in his steel mills, and the fact that newspapers published yearly lists of his dead and wounded employees that were as long as similar lists for battles in the American Civil War. End-of-life charity gives those who became rich by behaving poorly toward others the chance to salvage their reputations for posterity—calculating they can't keep earning money forever, knowing that they will one day have to leave it all behind. Today, over 200 of the ultra-rich around the world have signed the Giving Pledge, in which they promise to give away at least half their wealth to charity, either during their lives or after they die.* But without death, who cares about all that boring legacy stuff? Without death, where's the motivation for someone who's already decided it's fine to hoard billions to suddenly give their wealth away? After all, anyone who's not going to die still has to plan for the next several thousand years of their life, and all that money *could* come in handy. "You can't take it with you" doesn't carry much weight if you never have to go.

Worst of all, this immortality scheme is a medical procedure: it takes resources and money to perform and to maintain, and this means that not everyone on the planet will be able to afford it. This could very easily result in humanity breaking into two classes: the rich, who can afford to live forever, and the poor, who will live and die and be forgotten. This is an almost cartoonish dystopia of perpetual and self-reinforcing inequality, and even *insane* supervillains have standards.† Good thing there's an easy way to not just mitigate these horrible and society-destroying downsides, but

---

*That said, the Giving Pledge has no exact text, doesn't specify any specific causes that will be helped or how funds will be spent, is *specifically* not a legally binding contract, and carries no weight if broken.

†A great example of this is in 1997's "Batman/Captain America" comic, written and drawn by John Byrne, where the Joker ends his partnership with the Red Skull when he realizes he'd accidentally teamed up with an *actual Nazi*. (Joker: "You mean that's not just some crazy disguise?? I've been working with a Nazi?!?" Red Skull: "But of course. Why are you so upset, Joker? From what I have read of your exploits it seemed obvious you would make a superb Nazi!" Joker: "That mask must be cutting off the oxygen to your brain. I may be a criminal lunatic, but I'm an *American* criminal lunatic! Keep back, boys! This creep is mine!")

to actually avoid every single one of them with 100% certainty! Here is the secret:

Only you must become immortal.

If you are the only one who lives forever, then all the sociological implications disappear, because this no longer affects any civilizations: it's just a fun thing you're doing! Yes, it'll take guile and misdirection to ensure that all the researchers and medical technicians helping you never put the pieces together themselves and never succeed in duplicating your achievement in anyone else. But pull that off, and you win all the advantages of an endless life with none of the downsides. An immortal human can continue to learn forever, far beyond the limits imposed on the rest of us by biology. They can see the vast swath of human events and make connections that nobody else can see, becoming an expert in any number of fields of study that one human life simply isn't long enough to contain. They can rely on their youthful body to fight off infectious diseases that would kill someone more decrepit. Over dozens of lifetimes of study they can forget more than the rest of us will ever learn, becoming an enlightened leader, a god, living and learning and loving as they choose for as long as they want—*and humanity doesn't have to suffer for it.* Quite the opposite: they're *saving* humanity from suffering, in a real, measurable, and literal way, simply by withholding both this treatment and the knowledge of its existence from the rest of civilization. Their life may be lonely, filled with friends who enter it, grow old, and die in the comparative blink of an eye. They may be forced to move every few decades, keeping their secret by donning and shedding identities like layers of clothing. But this life is a long one, and it is theirs to keep for as long as they want it, for as long as they can hold their tongue.

You will save the world from war, from starvation and massive structural inequality, from civilizational collapse and disaster, and you will do it by ensuring—by swearing on all that you consider holy—that nobody else will ever use your technology. You will help the world by helping *only yourself.*

And that is the very definition of enlightened supervillainy.

POSSIBLE REPERCUSSIONS IF YOU'RE CAUGHT

What are they gonna do, kill you?

EXECUTIVE SUMMARY

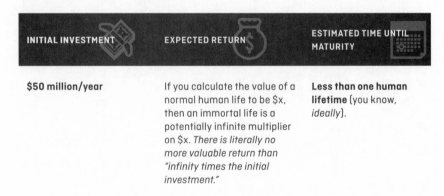

INITIAL INVESTMENT	EXPECTED RETURN	ESTIMATED TIME UNTIL MATURITY
$50 million/year	If you calculate the value of a normal human life to be $x, then an immortal life is a potentially infinite multiplier on $x. *There is literally no more valuable return than "infinity times the initial investment."*	**Less than one human lifetime** (you know, *ideally*).

ENSURING YOU ARE NEVER, EVER, EVER *FORGOTTEN*

And on the pedestal these words appear:
"My name is Ozymandias, king of kings:
Look on my works, ye Mighty, and despair!"
Nothing beside remains. Round the decay
Of that colossal wreck, boundless and bare
The lone and level sands stretch far away.

—*Percy Bysshe Shelley, "Ozymandias" (1818)*

We are now reaching the final section of this book, and you—assuming you have successfully pulled off all the capers I've described thus far—are reaching the peak of your glory. Your name is synonymous with greatness, genius, and grandeur. You are the most powerful and storied person to have ever lived, and you have accomplished more than anyone else has in the past or is ever likely to accomplish again. There is but one thing left that could possibly threaten you now . . .

. . . the possibility that your astounding exploits could one day be *forgotten.*

Sure, everyone on Earth knows you, but what of their children, and their children's children? Even with the best efforts of the previous chapter, there may still come a day when you die. When that happens, you will slip first into legend, then myth, then obscurity, and then into the endless and forgotten past where the untold billions dwell. Your legacy can—and will—be lost, unless you take the steps, now, to avoid it.

But rage not against this inevitable dying of the light, for you *can* avoid this fate! The Ozymandias in that opening quotation may have died and his kingdom may have fallen, but we're still talking about him millennia later, and you deserve nothing less. In fact, you deserve more. You deserve not just a few measly millennia, but epochs! *Eons.* You are going to ensure the story of your exploits lives as long as life remains on Earth, and then *longer* still, past the death of the planet and past the very end of the stars themselves. In this chapter, you're going to do all you can to ensure the universe will never forget the name . . .

[YOUR NAME HERE]

Look upon This Sidebar, Ye Mighty, and Despair

Ozymandias was what the Greeks called Ramses the Great, often regarded of the greatest of the Egyptian pharaohs, who ruled from 1279 to 1213 BCE, which itself was the greatest period for Ancient Egypt. His mummy was discovered in 1881 CE, but by 1976 was deteriorating: modern fungal growths had colonized it and were destroying the body. The mummy was flown from Cairo to France for preservation, where it was greeted with full military honors at Paris–Le Bourget Airport, including an air force contingent and the Garde Républicaine. Ramses was even given a modern Egyptian passport, possibly to help reduce the risk of French researchers keeping his body. He is both the only ruler of Ancient Egypt and the only mummy to have been issued a passport in the modern era: not bad for someone who's been dead for over 3,000 years.

When in the year 6000 you have a passport issued to either your living body (ideal) or corpse (less ideal, but still, better than most), you will have matched the standard he set.

In the pages that follow, we'll explore different strategies for information preservation on a logarithmic scale across eternity, starting with ensuring it survives 1 year, then 10, then 100, and so on. In most cases I assume whatever narrative you're preserving is a few paragraphs at most, but the expense of many of these techniques expands linearly with size, allowing the sufficiently funded supervillain to preserve anything from their autobiography, to their manifesto, to their combination autobiography/manifesto with full-color pin-up section in the middle.

Behold!

TIME PERIOD: AT LEAST 1 YEAR

SCHEME: POST IT TO THE INTERNET

COST: $0

LO

—the first message sent on ARPANET, the predecessor of the internet: the intended message was "LOGIN," but the system crashed before all the characters could be sent (1969)

There was a time when information put on the internet could be erased easily: it was stored only on the server you uploaded it to, storage and bandwidth were expensive, and the only impulses that existed to archive postings were either historical or altruistic. If nobody who read what you posted had made themselves a copy, you had a fair chance that when you deleted it, that information was actually being destroyed. It was great! You could type whatever you wanted and then deny it later and people just had to take your word for it!

That era ended by 1981. That's when Usenet discussion groups—an early form of internet public forums, launched just one year before—were first archived.* In 1996 it ended for the web (itself launched on top of the internet in 1991) when Brewster Kahle and Bruce Gilliat wrote a program

---

*There were Usenet archives before 1981, obviously, and some may be out there lurking on forgotten hard drives, but 1981 is the date of the earliest post in the largest publicly accessible Usenet archive that's survived to the present day: the DejaNews collection, purchased by Google in 2001.

designed to visit *every* website it could find and save a copy—and then do it *again* a few months later, so that changes over time could be recorded. When they publicly revealed the results of their work five years later, they called their service "The Wayback Machine," and today it contains over 20 years of web history and includes snapshots of over 439 billion web pages. Most of the pages that were available on the internet between 1991 and 1996 are now lost,* and the ones we have from that early era still exist only because the people behind the Wayback Machine started making copies of whatever they could grab without asking anyone for permission first. The fact that some branded Kahle and Gilliat as villains for this, chastising them as pirates for pulling off an audacious and ambitious scheme that these petty people couldn't yet fully understand but would one day thank them for, should feel deeply familiar to the experienced supervillain.

Anyway, this one's simple. To store something for free on the internet with the reasonable expectation that it will last at least a year, simply upload it to a free web-hosting service, or put it on your social media profile. Done. Easy.

A little . . . *too* easy.

In fact, the real challenge online is being forgotten. The information you post publicly is routinely copied into private databases of search engines, spammers, and corporations—and even if it wasn't, there's still lots of information being generated about you as you move throughout the internet, kept by people who have a financial interest in profiling who you are and what you do, usually because it means they can sell you something.

This means—just to pull an example out of the air—that you could click *just once* on an ad for sex-crazed dads in your area at 1:35 a.m. on September 23, 2022, and that fact could survive, somewhere, in some tracking database

---

*"Most" is an estimate here, since it's impossible to know what you've lost when you've never actually had a full index of what you had, but it's fair to say that most early versions of early websites are gone. Even the world's very first website—Sir Tim Berners-Lee's homepage at http://info.cern.ch/hypertext/WWW/TheProject.html, which went online on August 6, 1991—is a reconstruction, based on a snapshot saved in 1992.

for decades, possibly centuries. That one click could, in fact, be your true legacy: the one thing you created through your direct actions that lasts longer than anything else. The neuroscientist David Eagleman once said, "There are three deaths: The first is when the body ceases to function. The second is when the body is consigned to the grave. The third is that moment, sometime in the future, when your name is spoken for the last time."

We have now given to ourselves a fourth death: that moment when— even if it's unknown to you—that last fact about your life is deleted, lost, or corrupted. You will pass from this world for the final time when nobody knows that it was once late, and you were once tired, and bored, and curious about which particular local horny dads were online and eager.

So let's ensure *that* never happens by investigating longer ways of intentionally storing something much less embarrassing!

TIME PERIOD: AT LEAST 10 YEARS

SCHEME: POST IT TO WIKIPEDIA

COST: $0

History is written by the victorious Wikipedia editors.
—*Marya Hannun (2013)*

Founded in 2003, MySpace was, by 2006, not only the world's largest social networking site, but also the most visited website in America—beating out even Yahoo (more impressive then) and Google (still impressive now). But its star faded over time, and in 2019—after months of complaints from users who couldn't find their data—MySpace finally admitted that anything uploaded to their site more than three years ago was gone. They blamed an unnoticed error in a server move for the data loss, but many quickly noted that it's usually pretty hard to lose that much data without *anyone* noticing, and it would certainly save a struggling company a lot of money not to have to keep thirteen years of old mp3 files around anymore. I'll never forget you, gold_digger_cover_by_nickelback-2007_NOT_A_FAKE.mp3.

If MySpace had stored all their data someplace where anyone could access it (say, in a massively distributed database, where anyone at any time could download a complete copy for their own use) and had also allowed anyone to set up identical copies of their site—called "mirrors"—on servers around the world, then that data would not have been lost. But that's not how MySpace—or indeed, most private, for-profit websites—operate.

But that is how Wikipedia does things.

Destroy the original Wikipedia, and its mirrors will survive. Destroy all the mirrors and there are still countless copies of both its data and software stored offline all over the world, in hands both public and private. All these factors, plus its nonprofit and noncommercial nature, makes it extremely likely that the information on Wikipedia will still be around in a decade or more—which is more than can be said for many websites, personal servers, and even social media giants like MySpace.

So all you need to do now is get a Wikipedia page about yourself, put your manifesto on it, and you're set. Easy, right?

Well . . . not quite. There are some significant downsides to having an eponymous article, as Wikipedia will tell you itself on its straightforwardly titled "Wikipedia: An article about yourself isn't necessarily a good thing" page. Obviously, any information you add to your article could be changed or removed, and there's very little you can do about it. There's also the risk posed by lies, hoaxes, and vandalism—all of which have snuck into other articles and lasted for years before they're caught (see the sidebar for some all-time hall-of-famers). And even without lies, the truth could be even worse! All the embarrassments and losses you'd most like to forget, once they become public, are going to be added to and even highlighted within your Wikipedia page: the "controversies" section is always the juiciest part of anyone's article. And since Wikipedia's authority with search engines means their article about you is often the first thing anyone sees when they look you up, this means that for the rest of your life, anything you post online will likely *always* be second in search results to something that others can alter and which you absolutely cannot control.

Wikipedia's Greatest Hoaxes

Wikipedia actually keeps track of how long hoaxes last on their site on a page titled "List of hoaxes on Wikipedia." The longest one they've caught (so far) was an imaginary Egyptian scholar whose article lasted 14 years and 10 months in the encyclopedia before someone noticed that there were no other records of him and that his name, Sheikh Urbuti, was a reference either to Frank Zappa's 1979 album of the same name or to KC and the Sunshine Band's 1976 disco/funk hit "(Shake, Shake, Shake) Shake Your Booty." "Bont," a made-up ball game from France that crossed croquet with hockey, lasted 12 years and 8 months before being caught. And for 5 years and 3 months, Wikipedia claimed on its page for high fives that in the rebooted Planet of the Apes movie series, war began when an ape was offered a high five that was "down low," only to have the offered hand quickly removed, and the ape—infuriatingly, one imagines—deemed "too slow."

But worst of all, for our purposes, is the fact that Wikipedia maintains notability standards for the subjects of all of its articles. Not everyone will have an entry in Wikipedia, and it is up to the Wikipedia community to decide that you are worthy of notice. Create a page for yourself without meeting their notability standards, and that page is likely to be deleted. Since only administrators can view deleted articles, this derails the entire purpose of this endeavor, and for the vast majority of the world, your words will disappear.* This would seem to be the final nail in the coffin for the "store a message on Wikipedia" idea . . .

. . . if Wikipedia didn't store complete *revisions* of all its active articles.

With every edit you make to an article, the old version is stored on Wikipedia's servers, where it can be viewed, linked to, and shared online. Even better, these older versions of articles are *also* included as part of the complete-text Wikipedia download, which contains every revision of every page. Vandalize a popular article with your message and then revert it, and your first version of that article will remain, online and available to all who seek it out, likely for as long as Wikipedia lasts. It won't be highlighted on Wikipedia, of course, but it will be there, stored and accessible to anyone online, mirrored on servers around the world and on the personal computers of everyone who ever downloaded the complete text of Wikipedia.†

I recommend the article about chickens: 25.9 billion of them lived on Earth in 2019, which is an order of magnitude more than any other bird and enough to count three chickens for every human. As chickens have had a major effect on medicine, science, history, society, agriculture, and human diets worldwide, it is my long-standing and considered opinion that it's *extremely* likely the chicken article will still be on Wikipedia ten years from now.

And yes, while it is generally considered villainous to vandalize Wiki-

---

*Administrators are a small curated tier of Wikipedia editors. As of May 2021, there were only 1,097 administrators on the English version of the site, compared to the 41 million registered accounts, of which 140,000 are actively editing pages.

†Note that Wikipedia administrators do have the authority to execute what they call a "Revision-Delete," introduced in 2010, which *does* hide your revision from public view. But don't worry: this is a tool with strict guidelines, intended to be deployed against edits that contain blatant violations of copyright, are grossly offensive or degrading, contain harassment or threats (so keep them on the down-low), include malicious code, contain purely disruptive material, and so on. Keep your message relatively short and not blatantly offensive and it's unlikely you'll fall into the clutches of a RevisionDelete.

pedia for your own benefit, irritating the selfless volunteers who are simply trying to collect, organize, edit, and distribute all of human knowledge for free, *isn't that what we're all here for?*

VILLIPEDIA

No regrets.

TIME PERIOD: AT LEAST 100 YEARS

SCHEME: PUBLISH IT IN A BOOK

COST: IF YOU'RE LUCKY, THIS ONE CAN ACTUALLY MAKE YOU MONEY

Every good story needs a villain. But the best villains are the ones you secretly like.

—Stephanie Garber (2018)

The longer the time period, the less likely it is that any one message will survive. Mitigate that with publishing, a worldwide industry whose whole purpose is to create multiple copies of the same text, package them conveniently, and distribute them for storage in homes, libraries, and bookstores around the world! Publishing functions as a massively parallel backup scheme for information, there are (ideally) cultural taboos around burning books, so they tend to stick around, and, best of all, it's even sometimes profitable.*

"But," I can hear you saying, "we already did that distributed backup thing with Wikipedia in the last section! Why do we have to do actual publishing now?" And the answer is this: computer data degrades too, and it

---

*Thanks for purchasing this book. And if you didn't buy it, thanks for taking it out from the library or borrowing it from a friend, who themselves purchased it! *There are no other viable ways in which you can obtain this book, so let's all stop speculating about them right now.*

can do so in faster and more surprising ways than older media. Obviously, data can degrade physically as the media it's stored on fails: hard drives crash, magnetic media degrades, and tape drives get chewed up. But because computer storage mediums are constantly evolving, in living memory mainstream computers have gone from stored data on punch cards, to magnetic cores, to magnetic tape, to floppy disks of various incompatible sizes, to optical disks of various incompatible standards, to flash drives with no moving parts, and none of these technologies are compatible with the others. For old data to survive, it must be continually moved from one storage medium to another. Neglect that, and something called "bit rot" sets in: the metaphor is the data rots, but even if the medium survives, the data itself is still frozen in time as the world around it continues to evolve. Hardware to read the data stops being produced, file formats get forgotten, and within a few short years you can end up with media that nobody remembers how to access anymore.

It's happened before. There were more than 3,000 photos gathered from the surface of Mars in 1976 by NASA's Viking probes that had never been processed. When recovery was attempted in 1988, the data could still be found, the medium it was stored on had not decayed, and the programs to interpret it still existed—jackpot, right? But the catch was that the *hardware* that those programs needed to run on was long gone, along with the software's source code, which meant the software couldn't run on current machines and had to be rebuilt. Recovering those images just *12 years* after they were taken required reverse-engineering the data format from scratch.

And that's not even the only case within NASA itself! When tapes containing images taken by the Lunar Orbiter spacecraft were going to be scrapped in 1986, Jet Propulsion Laboratory archivist Nancy Evans decided to try to preserve them instead of destroying them. But she soon discovered that the hardware required to read the tapes—heavy, refrigerator-sized Ampex FR-900 tape drives, manufactured exclusively for the U.S. government—hadn't been produced for 25 years. It took until 2008 for four broken drives to be located, restored, cannibalized for parts, and used to create a single working machine that could recover the images. And efforts like that only work if the data is still good: of the millions of old data tapes NASA has, hundreds of thousands have experienced what the organization itself describes as "deplorable conditions." Many of them are so fragile that even reading them causes the recording medium itself to flake off and disintegrate, meaning that if the data they contain isn't *already* lost forever, if it's not copied to some new medium the next time it's used, it will be.

In other words: you can save data on a computer, but it can still decay, and you might not know you've lost it until you try to access it. Some have worried this invisible decay combined with information's perception of invulnerability will lead to vast amounts of it being lost: a digital dark age. There's precedent here for that too: at least 75% of all silent films ever made have been lost, something the Library of Congress described as "an alarming and irretrievable loss to our nation's cultural record."* Remember floppy

---

*Most of these movies were produced on nitrate film, a highly flammable medium that, if it doesn't catch on fire, still deteriorates over time. And the valuable silver used in the production

disks, the once-ubiquitous, now-obsolete medium whose major lasting impact is the inspiration for the otherwise-inscrutable "save" icon? The data you saved on them lasted around 10 years before it started to decay. Anything saved on once-even-more-ubiquitous CD-ROMs is expected to last 100 years at most, and that's under ideal conditions. And the lifetime of the data on a modern flash drive is anywhere from 1 to 100 years, depending on the quality of the drive and the temperature at which it's stored.

It's true that we may get better at preserving our data in the future. But it's also true that there's not a single book published 12 years ago that we could read then but struggle to understand now. There is no collection of 100,000 books that could disappear as easily and unnoticed as 100,000 files on a forgotten hard drive can. If you want your data to survive the long term, you need to rely on a technology that's been produced for thousands of years. You need writing: actual, physical, analog writing. *You need to publish a book.*

But the downside of being a supervillain is being aware of how often there's some meddling little detail that wants to foil our schemes, and here—the challenges of finding a literary agent willing to represent a supervillain aside—our problem is this: *language evolves.* 50-year-old books can sometimes seem a little dated. Books that are 100 years old can be a little challenging because of words, phrases, references, idioms, and metaphors that no longer carry cultural currency. And while Shakespeare's plays aren't yet even 500 years old, they're routinely published today in a format with his original words on one side and modern explanations, translations, and historical context on the facing page. That is both wild and worth repeating: *the greatest plays in English history are no longer accessible to the average speaker of English.* And *Beowulf,* published in Old English (then called "English") somewhere between 700 and 1100 CE, is even more inscrutable: its first line reads "Hwæt. We Gardena in geardagum,

---

of early film could be recovered and sold by recycling the film: another motivation not to keep it around.

þeodcyninga, þrym gefrunon, hu ða æþelingas ellen fremedon."* Languages change over time, with the original eventually becoming unintelligible to modern ears, and that's going to be a problem no matter what we do.

Well . . . *forsooth.*

Are Shakespeare's Plays *Truly* the Greatest in the English Language?

Shakespeare scholars sure do think so! But I've actually read some of them, so I speak with authority when I say that his plays are okay, *I guess*, but it's hard to argue they couldn't ever be improved upon. For example, did you know that not once in Shakespeare's works does even a *single* character gain access to a giant robot suit, much less employ it to lay waste to their enemies? Academics will argue that their beloved bard captures the very heart of the human condition with sublime nuance and rapturous magnificence, but any conception of humanity that excludes the ever-present desire to possess a robot large enough to climb inside, and which also fires lasers out of its eyes and missiles out of its hands, is one that feels somewhat blinkered.

*A modern translation is something close to "Listen: we all know of the mighty Spear-Danes in the old days, and how they were glorious and courageous warriors."

Good thing that, like all great villains, we have an unexpected and heretofore-unrevealed weapon that will turn the tide of battle! What we're going to exploit now is the fact that historical linguistic analysis shows *not all words evolve at the same rate*. While some can change in a matter of years (the term "computer" once meant "someone, usually female, who performs calculations by hand": a meaning lost only a few years later when we invented, well, computers), others can remain stable for millennia. Here's our secret weapon: generally, the *more* frequently a word is used in everyday language, the *less* likely it is to change over time. And how can we know what words will be frequently used in the future? We look to the past, where we discover that most stable words in history tend to be short and simple ones reflecting core concepts that are encountered frequently throughout human experience: things like colors (think "red and white" over "coquelicot and alabaster"), basic numbers ("one," "two"), body parts ("eye," "hand," "ear"), simple actions ("come," "see") and common sights ("mountain," "star," "fish," and so on). It's estimated that only 6% of these words will change every 1,000 years, though some estimates put it as high as 14% or as low as 4%—the language being spoken can influence these numbers.

So one way to help ensure your message remains intelligible in the future is to limit yourself to the common words reflecting these core concepts. Here are the 40 most stable words across languages described in a 2008 computational research paper,* listed in alphabetical order.

> blood, bone, breasts, come, die, dog, drink, ear, eye, fire, fish, full, hand, hear, horn, I, knee, leaf, liver, louse, mountain, name, new, night, nose, one, path, person, see, skin, star, stone, sun, tongue, tooth, tree, two, water, we, you

A threat that has a good chance of being understood 1,000 years from now is "I see you. I come. Nose sees knee. Skin sees blood. Bone sees sun. You die."

---

*It's titled "Explorations in Automated Language Classification" and it's in the bibliography, if you're interested!

The fact that "die" is on that list helps us, but that's still a pretty limited set of words. But we can increase our expressive potential here a lot without significantly sacrificing intelligibility: since we're writing in English, we can add things like more of the common English pronouns ("she," "he," "they," "who," "my," "your"), conjunctions ("and," "but," "so"), auxiliary verbs ("am," "be," "do," "have," "will," "may," "should"), articles ("a," "the"), and prepositions ("on," "of," "for," "in"), all of which are commonly used and unlikely to change. And at the risk of slightly decreasing linguistic stability, we can also include some more of the very common English words,* which gives us the following list of Words to Be Used in Temporally Stable Threats:

a, all, am, and, ash, bark, be, belly, big, bird, bite, black, blood, bone, breasts, burn, but, claw, cloud, cold, come, could, die, do, dog, drink, dry, ear, earth, eat, egg, eye, feather, fire, fish, flesh, fly, foot, for, full, give, good, grease, green, hair, hand, have, he, head, hear, heart, hers, his, horn, hot, I, in, knee, know, leaf, lie, liver, long, louse, man, many, may, moon, mountain, mouth, must, my, name, neck, new, night, nose, not, of, on, one, path, person, rain, red, root, round, sand, say, see, seed, she, should, sit, skin, sleep, small, smoke, so, stand, star, stone, sun, swim, tail, that, the, their, they, this, tongue, tooth, tree, two, walk, water, we, what, white, who, who, will, woman, yellow, you, your

A more nuanced threat that still has a good chance of being understood 1,000 years from now is "I am a big person of fire and smoke. You will know my name. I see your men and women. I see their skin, full of hot blood. I claw at them, and their blood rains on the earth. They give me good flesh. Their hearts know cold. In one night, your men will die. Your woman will die. And you will die, on your knees, my big foot on your small head, my name flying from your mouth: [HI, READER OF THIS BOOK, YOU'RE SUPPOSED TO PUT YOUR NAME HERE]."

---

*These words are substantially based on the lists developed by linguist Morris Swadesh, first published in the early 1970s. His goal was to nail down a set of basic human concepts that could then be used to compare languages to one another.

Not bad. There's just one more thing you should do now: make sure your message is translated into multiple languages, all bound together in every copy of your book. Remember, the Rosetta Stone is famous not for what it says but for how it says it: it has the same message engraved three times, in Ancient Greek, Egyptian hieroglyphics, and Demotic script. When it was found in 1799 CE, Ancient Egyptian hieroglyphics hadn't been understood by anyone living on Earth for over 1,200 years. But afterward, our existing knowledge of Ancient Greek—which we had already managed to decipher—was the key to understanding the other two languages, which in turn unlocked a wealth of knowledge of Ancient Egyptian culture and civilization. By doing the same and printing multiple translations of your text in the same book, you greatly increase the odds it will be understood by those living hundreds of years from now.

The three most common languages on the planet in 2021 are English, Mandarin Chinese, and Hindi, with 1.3 billion, 1.1 billion, and 600 million speakers, respectively. Use these languages in your book and you'll be off to a great start! (书里用了这些语言，便有了金彩的开场白!) (अपनी किताब में इन भाषाओं का इस्तेमाल करें, आपकी शुरुआत अच्छी होगी!)

What Does the Rosetta Stone Actually Say?

For all its impact on our understanding of history, the Rosetta Stone is actually pretty dull reading. It describes a decree, passed by a council of priests in 196 BCE, that celebrates the anniversary of the coronation of their boy-king Ptolemy V. It then lists some of his accomplishments (their boy cut taxes), ignores his setbacks, and then pledges that some new statues will be constructed in his honor and that his shrines will be decorated with ten golden crowns each, with private citizens encouraged to make their own similar shrines at home. But—most useful to us—it also decrees that copies of itself be placed in temples throughout Egypt, translated into three languages: Egyptian hieroglyphics (the writing of the gods: at that point hieroglyphics were a dead script understood only by the priesthood), Demotic script (the writing of documents: Demotic was the Egyptian script used by current administrators), and Greek (the writing of the people). Thanks, boy-king Ptolemy V!

To sum up: print as many copies of your book as you can, and you increase the chances it survives. Use simple, stable words, and you increase the chances it'll be understood. Bind multiple translations of your text together, and you're vastly increasing the chances your words will be read even if the English language *itself* is forgotten. And finally, if you make your book interesting or notorious enough, foolish mortals (and genius supervillains) will pay for the privilege of owning and preserving your legacy. Thank you, by the way.

And *that* is how you write a book to last a hundred years or more.

TIME PERIOD: AT LEAST 1,000 YEARS

SCHEME: HIDE IT FROM HISTORY

COST: $18,900 USD A POP, PLUS TRAVEL EXPENSES

Every breath you take is a step toward death.

—*Ali ibn Abi Talib (600s CE)*

The book you printed in the last section could last a thousand-plus years if it's preserved properly—most books can. In fact, most *materials* can. We have ancient erotica from 79 CE, hand-cranked computational machinery from around 200 BCE, papyrus with writing on it from 2500 BCE, leather shoes from around 3500 BCE, and even preserved human bodies (in mummy form from around 7000 BCE and in bog-person form from around 8000 BCE).

Tell Me About That Cool Old Stuff You Mentioned

Gladly!

The **erotica** refers to art recovered from Pompeii, the city destroyed and buried in the 79 CE eruption of Mount Vesuvius.

There's older erotica too, of course, but the Pompeii ones are memorable because of how abundant and well preserved it is. The Pompeiians were much more open about sex than those who excavated their remains in the 1800s were, and when this art was discovered, it was scandalous. Imagery of erect penises was ubiquitous, used in everything from sculpture to paintings to wind chimes to oil lamps. (One painting of the god Priapus with an enormous penis was so shocking that it was plastered over and forgotten, only rediscovered after rain damaged the plaster covering in 1998.) In 1821, the King of Naples ordered all this penis and vagina and sex stuff to a semi-private room in the Naples National Archeological Museum, and even that wasn't enough to stop the raw sexual power of these ancient naked people: the doorway to this "secret museum" was bricked up in 1849. Since then, it has been closed and reopened several times, most recently in 2000 CE.

The **ancient computer** is the Antikythera mechanism, a complex piece of hand-cranked Ancient Greek clockwork that could predict eclipses, model the movement of the moon, and, if that wasn't enough, *also* indicate when the next Olympic Games were going to be held. Discovered in 1901 CE, it took until 2006 CE for us to reconstruct how it worked, helped in part by X-ray scans that revealed more of the mechanism and some explanatory inscriptions still legible inside it. Clocks of a similar complexity wouldn't be made again until the 1300s CE in Europe, and as those medieval clockmakers worked to create their masterpieces, they never knew that this technology had already been achieved over a thousand years ago and was there, waiting to be rediscovered, in a shipwreck at the bottom of the Aegean Sea.

The **papyrus** are logistical documents (found in a hot, dry cave in the Egyptian desert in 2013 CE) written by a man named Merer (meaning "Beloved"). Merer and his team helped near the end of the construction of the Great Pyramid of Giza, and his diary describes moving the white limestone (used for its outer casing) from where it was quarried in Tura, and taking it down the Nile to Giza. More on that pyramid later!

The **shoe**, known as the Areni-1 shoe, is (currently) the world's oldest leather footwear. It was found in excellent condition in 2008 in a cool and dry Armenian cave and is a woman's U.S. size 7 (UK size 5).

And the **mummies** are the Chinchorro mummies: the oldest mummified corpses in the world! To date, 282 of them have been discovered in the Atacama Desert in Chile, where its hot and dry environment encouraged bodies to desiccate before they decayed, and high salt levels in the sand discouraged bacteria from growing. This ensured that the simple act of burying a dead body could, in some cases, result in its mummification and preservation. Bog people are mummies preserved in a different way: there, the acidic water and low-oxygen conditions of peat bogs can tan and preserve skin while softening bones. This results in the "deflated sac" appearance of some bog bodies (no judgment here).

At this time scale, what matters is less the material your message was printed on and more the conditions in which it was stored. Stone weathers away in, well, weather, so you'll want someplace warm and dry that will keep this material out of the way of water for a good long time, like a desert cave. For organic matter, anything that arrests bacterial growth helps: dry, salty sands for mummies, and acidic, unoxygenated water for bog people. And yes, if right now you're thinking that getting your message tattooed onto your skin before diving into a peat bog is a viable technique for communicating with the distant future: you're right, it is. You could absolutely do that.

But here's the catch. Although we recognize that this goes against every instinct for glory a supervillain has, you must first allow whatever object you choose to carry your wonderful words, the very culmination of all you have struggled for in your lifetime and/or this chapter . . . to be *forgotten*.

Every item we mentioned at the start of this section—each of them a unique and priceless historical treasure—didn't make it to our era on purpose, passed down through the ages by an unbroken line of careful archivists intent on preserving them for future posterity. We have them because they were lost—disregarded and forgotten in some environment that happened to keep them around, safely removed from that destructive sequence of events we're all constantly experiencing called "history"—and then, recently, we found them.

How much of an advantage does hiding your artifact from the rest of humanity provide? We can look to the Seven Wonders of the Ancient World for comparison. These were crowning achievements for the civilizations who produced them, public symbols of their prestige and power, and things that they'd absolutely want to preserve: the Hanging Gardens of Babylon (beautiful garden), the Great Pyramid of Giza (giant tomb), the Colossus of Rhodes (giant statue), the Lighthouse of Alexandria (giant lighthouse), the Mausoleum at Halicarnassus (giant tomb), the Temple of Artemis (giant temple), and the Statue of Zeus (giant statue, but *this* one's sitting on a chair). Today, *all but one of them is gone*—the only exception being the pyramid, whose white limestone exterior has long since been harvested for other projects but whose damaged and looted remains still stand. These great achievements of humanity were felled by plundering, by arson, by earthquakes, and by decay. Even once-treasured cultural artifacts can die from neglect: the Great Library of Alexandria famously burned in 48 BCE, but what's less well known is that it survived that fire, only to be done in by centuries of declining support and prestige until it disappeared from the historical record entirely around 260 CE.

Artist's conception of the fall of the Library of Alexandria. Did you know that artists can conceive of whatever they want, and there's no law that says it has to be historically accurate? This artist shows the ancient Egyptian library with modern English signage, bound books, and a dog wearing a backward and entirely anachronistic baseball cap! Nice try, artist. Conceive again, artist!!

Over a long enough time frame, even the most prestigious artifacts can become unpopular, or politicized, or a liability, and anything that requires upkeep is eventually going to encounter a generation that doesn't care to pay for it (like we saw with chantry in Chapter 8). Even innocuous objects can be lost if a culture decides it hates you enough. When Spanish invaders and Catholic priests tried to burn every Mayan book in existence in the mid-1500s CE, they succeeded, destroying all but *four* of them. But those four books that survived did so because they had something in common with all the other objects mentioned here: they were small enough to be hidden, sturdy enough to remain stable, and lucky enough to escape the generations of humans who'd want to burn then, or smash them, or simply do a bit of tidying up. And eventually, we modern people were lucky enough to find them again.

There's that word again: "luck." Unfortunately, at this time scale, there are no guarantees, and there is always some measure of good fortune required for success. You must hide your message-carrying object well enough that it can sidestep thousands of years of human events, but not *so* well that it's never found again. The place you choose to hide it must not only have conditions that will preserve your object now, but those conditions must remain the same for a millennium or more. You can—and should—increase your odds by hiding multiple copies of your object in different locations around the world, but when you're trying to talk to the people living a thousand years in the future, there are no guarantees, and none of us know what the future will hold.* I can't promise anything.

But I *can* show you how to unfairly tilt the odds in your favor.

First is the environment: you want to store your object somewhere humans don't normally poke around—so this excludes most of the surface world and anywhere underground that a meddling human with a shovel could prematurely reach—but this environment should not be so inaccessible that it can't be stumbled across eventually. And as for the material . . . well, let's speculate for a moment. If we put aside all practical considerations and instead imagine what an *ideal* material for our purposes would be, we would say that your object should be made out of something so hard that it can keep its shape for millennia, even if exposed to the crushing weight of soil or the weathering effects of water. It should obviously resist decay. And it should also be *beautiful*, because it is carrying your beautiful words, and because it should attract the eye both when it's spotted and when it's displayed in whatever museum or place of worship it ends up in. And while we're making a wish list, let's also demand that this material somehow acts to preserve *itself*, automatically generating a hard protective coating whenever it's exposed to the elements so that whatever's inside is saved.

---

*Though remember to check Chapter 6 if you'd like to change that.

Lucky for you, both this environment and material exist. And lo, you shall exploit them both: first by casting thy message in bronze, and second, by chucking thy message into the sea.

Bronze is an alloy of mostly copper (around 88%) with tin making up the rest. It's hard, it's dense, and it can last a long time. The oldest bronze sculpture that made it to the modern era is around 4,000 years old: it's from the Indus Valley civilization, cast somewhere between 2300 BCE and 1750 BCE, and today we call it *Dancing Girl*. It depicts a naked woman (please note: history is sometimes naked) with long arms, shown standing with one hand on her hip. The information contained in this statue (including, but not limited to, the fact that nude women sure did exist in the past) has survived around 4,000 years so far, which is at least 40 times longer than the person who cast it, and at least twice as long as the civilization that person emerged from. If this sculpture had been of triple-translated words instead of a naked lady, we'd probably know even more about the people and culture that produced it, beyond the fact that some of them thought naked ladies were neat.

While bronze is a great medium to get information into the future, it's unfortunately one that's also both valuable *and* recyclable: it melts at around just 950°C, a lower temperature than steel and one that's well within the reach of the furnaces of now-in-retrospect-obviously-named Bronze Age civilizations. For a bronze statue to survive, the bronze it's made out of has to reliably and consistently be considered more valuable in its "sit there and do nothing except be occasionally looked at" configuration than in any of the other shapes it could be relatively easily reforged into, including cannons, coins, cymbals, shields, weapons, nails, drums, bells, cookware, newer and less-dated statues, and more. Because of this, only a few full-sized bronze statues from Ancient Greece survived to the present day, despite them being considered the highest form of sculpture at the time. (Most of what we know about life-sized Greek bronze statuary actually comes from stone copies made by the Romans.)

That's where the ocean comes in: it keeps everyone's grubby hands off your bronzework, obscuring its location and making messing with *your* bronze much more difficult and expensive (and therefore uneconomical) than messing with anyone else's. Two of the few full-sized Greek bronze sculptures we *do* have, known as the Riace bronzes, were found underwater off the coast of Italy. They show two different life-sized naked men (again, history: occasionally nude) with impossible superhero-esque physiques, and were created between 460 and 420 BCE before being dumped into the sea at some point afterward.* They remained there for thousands of years, until they were discovered on August 16, 1972, by one Stefano Mariottini, who was spearfishing in the Ionian Sea for his summer vacation. Mariottini found the statues resting in less than 10m of water, 200m off the coast of Italy: the shape of a human hand was sticking out from the sandy ocean floor; he initially thought he'd come across a dead body.

The reason that hand was in such good shape was because bronze doesn't oxidize like other metals. When metals oxidize—and most do†— that simply means that they're undergoing a chemical reaction with oxygen found in both air and water, which generally causes them to corrode. You've seen this before when rust forms on a car: rust is just iron oxidizing, and over a long-enough time frame, anything iron can rust away entirely.

But bronze has a superpower.

When bronze oxidizes, it forms a thin, hard layer of oxides called a "patina" around its surface that actually protects the bronze underneath from further corrosion. *This* is what makes bronze such a great medium to send information into the future. When the Riace bronzes were discovered, the patina covering them was thin and detailed enough that features were

---

*It's assumed they were lost in a shipwreck rather than tossed intentionally, but no wreckage has been found.

†There are a few metals that don't oxidize at all! These unreactive metals are called noble metals, and while you may know some of them (silver, gold, platinum, and palladium), others may be less familiar (ruthenium, rhodium, iridium, and osmium). Unfortunately, while they're all beautiful and shiny candidates for our statues, they're also all either too soft, too malleable, too brittle, and/or too rare and expensive for our purposes of long-term information storage. Bronze hits the sweet spot between longevity, beauty, and price.

still clearly visible: the shape of their faces, the bushiness of their beards, and even individual muscles. Despite the millennia between creation and rediscovery, it was still clear that these were statues of handsome naked dudes. And when the figures were restored and their outer layer removed, their original fine detail remained, as clear as the day they were cast. The biggest loss sustained in the over 2,000 years they spent underwater was just one eye on one statue: instead of bronze, the eyes were made separately out of white calcite, and at some point, one of them detached and was lost. The rest of these statues made it to the future just fine.

Your statue can do the same. Like the Riace bronzes, you're going to cast your message in similarly scaled bronze: 2m long, half a meter tall, and a third of a meter wide. To give yourself an even better chance against corrosion, you're going to write using letters that are 3cm tall, raised from the surface by at least 3cm too: large enough to be read easily, even under corrosion. With 3cm letters and 1cm of space between them (and with 1cm of space between each line), that leaves room for 12 lines of 50 characters on both large faces, for 1,200 characters in total—if you're wondering what that looks like, it's exactly four times the length of the sentence you're reading now. And finally, you're going to leave your statue someplace where it can remain hidden from looters but still be located by chance on a long-enough timeline: off the coast somewhere, in water that's deep but not inaccessible . . . including but not limited to water 10m deep, some 200m off the coast.

A custom bronze statue as described here would cost around $18,900 USD to create today.* As bronze casting relies on making a mold of the statue, which can then be reused to make copies, any duplicate statues both cost less *and* increase the odds one of them will be found in the time frame you want. If you commission 500 of them with a 5% bulk discount—you

---

*This price was supplied to me by Matt Glenn, president of Big Statues LLC. Glenn is an artist specializing in bronze sculpture who was kind enough to entertain my extremely suspicious commission request without asking too many nosy, inconvenient questions. He may also entertain yours!

may be able to achieve a greater one, depending on any supervillainous negotiating abilities you possess—that means it costs just about $9 million to successfully deploy this scheme. Add on a generous $5 million more to cover the costs of quietly shipping them around the world and clandestinely dumping them off the coasts of all the world's oceans and seas, you're looking at less than $15 million to ensure, as much as one can, that your message will be found, read, and respected / feared / worshipped / canonized in the millennia to come.

But remember: that's just to increase your odds of success. If you haven't yet become a multimillionaire, one statue still buys you a ticket for this thousand-year memorial lottery, all for less than the price of a new car.

TIME PERIOD: AT LEAST 10,000 YEARS

SCHEME: BUILD A NEW AND BETTER MOUNT RUSHMORE WITH YOUR OWN FACE AND WORDS

COST: $64 MILLION USD FOR THE ENTRY-LEVEL CARVED MOUNTAIN, EXTRA TO INCLUDE THE SITE ALTERATIONS

God has given you one face, and you make yourselves another.
—*William Shakespeare (1599)*

When you're staring down 10,000 years, small and hidden things can too easily become lost-and-gone-forever things. Your best odds now come from going in the opposite direction, building something big, recognizable, and (as much as you can make it) permanent. Your best odds come from building your own bespoke *Mount Rushmore*.

Mount Rushmore, being a bunch of colossal heads carved into the side of a mountain, is already *deep* into supervillain aesthetics: all that's really missing is for it to be located on an uncharted island somewhere in the Pacific. And the sculpture has a real-life evil origin story too! The Black Hills the sculptures are carved into are sacred ground, promised by the U.S. government to the Lakota people in perpetuity by an 1868 treaty. But that treaty was respected only briefly: the Americans soon found gold in the Black Hills, which led to the U.S. government attempting to take the area back by force during the Black Hills War of 1876. When that war failed, the government instead worked to starve the Lakota people until they finally ceded the territory in 1877. As for the heads carved into the mountain, they were designed and sculpted by Gutzon Borglum, a white supremacist and *council member of the Ku Klux Klan*,* who overrode the original idea of including non-white figures like Sacajawea, Crazy Horse, and even Lakota leader Red Cloud and instead decided that the only people featured would be white male American presidents. Finally, the mountains themselves were renamed from the Lakota name—Six Grandfathers—after a lawyer named Charles Rushmore made the largest single donation to support Borglum's work.

All this to say: when it comes to villainous megaprojects, *there's definitely precedent.* And this one was even mostly completed! Four giant faces *were* carved into that mountain. The heads—originally planned to have torsos, but the money ran out—stand around 18m tall, and creating them took around 400 workers 14 years, from 1927 until 1941. The total cost for the project was precisely $989,992.32 USD then, which is closer to $18 million in today's currency. But don't start writing a check for your own mountain just yet, because megaprojects were cheaper then. A 2017 estimate in *The Washington Post* of what it would cost now to add then-president Donald Trump's face to the mountain† calculated that it would take 180 people about four years to add the figure. This included a staff of 25 designers (paid $100 an hour), 30 stone workers (paid $50 an hour), and 125 laborers (at $30 an hour). In total, the estimate came in at around $64 million for adding just one face.

---

*There's no evidence that Borglum ever *officially* joined the Klan, which let him publicly deny being a member while privately supporting them. He wrote in a letter to a friend that "While I am under no obligation to [the Klan] as an organization, I am under the obligation of deep friendship to many of their most prominent leaders and would do anything I could, publicly or privately, to serve them," and told another that "[The Klan] are a fine lot of fellows as far as I can learn and if they elect the next President, by gosh, I'm going to join 'em." Borglum served on the Klan Kloncilium, a committee whose responsibilities included drafting the Klan's national political platform.

†2017 was a weird time.

But look at what that $64 million gets you! The South Dakotan granite those faces are carved into naturally erodes about 2.5 cm every 10,000 years, which means that for every 2.5 centimeters deeper you carve your letters, you buy yourself another 10,000 years of legibility. At those rates, the Mount Rushmore noses alone would last 2.4 million years before eroding completely, and there would still be at least some evidence of the faces left as far as 7.2 million years in the future. What's more, Borglum had initially planned to have a massive inscription carved into the site with letters almost 2.5 meters tall, and deep enough that—if he hadn't started running out of suitable sites and moved Lincoln's head there instead—the words would've lasted an estimated 100,000 years before they completely eroded away. We can compare these estimates to the Sphinx in Egypt— also carved in place out of bedrock—and find that despite enduring 4,500 years of weathering and human contact so far, it's still recognizably a face, despite being carved out of softer limestone.[*] And of course, we've already seen how the Great Pyramid of Giza still stands, weathered, looted, but still world-famous.

The Georgia Guidestones

If you don't have $64 million for your own Mount Rushmore, you *could* always go with a smaller monument. The Georgia Guidestones—commissioned anonymously in 1979 and unveiled in 1980 in (you guessed it) Georgia, USA—are five

---

*Of course, the Sphinx is much less contentious than "a bunch of U.S. presidents, two of them slaveholders, carved into a disputed mountain by a council member of the KKK" is, which means Mount Rushmore might, depending on how people react to it in the future, last for a shorter time than the Sphinx does.

granite slabs arranged with a capstone on top and reach almost 6m in height. They're engraved with the same message in eight different languages across them: Arabic, Chinese, English, Hebrew, Hindi, Russian, Spanish, and Swahili.

Unfortunately, the message itself is pretty useless: it contains just ten instructions, including "Avoid petty laws and useless officials" (thanks for the tip, giant stones!), "Unite humanity with a living new language" (there's nothing wrong with multiple languages, as they're not fungible and actually help capture the depth and breadth of human experience and culture, giant stones!), and "Guide reproduction wisely—improving fitness and diversity" (why'd you have to bring eugenics into this, giant stones?). The stones have survived the handful of decades since their installation, though their small scale and accessibility have led to them being repeatedly vandalized.

But it's not just simple erosion that can do you in. The Six Grandfathers are in South Dakota, where temperatures drop below freezing. Any cracks where water can pool risks that very water freezing in winter, which then expands, weakening and damaging the sculpture much faster than erosion alone. To mitigate this, all cracks in Mount Rushmore are filled with silicon to prevent water pooling. But you're not going to bet your future posterity on suckers maintaining your sculpture for thousands of years, so what are your options?

One is to create it someplace where temperatures rarely, if ever, drop below freezing, like the Sphynx in Egypt. Unfortunately, the challenge at this 10,000-year time scale is that we can't be certain that any favorable climate will *stay* favorable. For example, the part of Africa that makes up the Sahara Desert today was, 6,000 years ago, covered in lakes and trees. And in Europe, from the 1300s CE to the 1800s, the colder summers and winters produced by the Little Ice Age resulted in crop failures, starvation, *and* rock-weakening freezing where there had previously been none. Human-generated greenhouse gases are likely to warm the planet for the next several thousand years, and that does reduce the odds of warm areas on Earth becoming colder (though see Chapter 4 for a possible mitigation technique), but at these time scales, it's impossible to say for sure. And given the size of these monuments, constructing several backup copies like we did with our bronze statuary isn't exactly feasible. Instead, all you can do is choose someplace warm, carve your granite slowly and carefully to avoid cracking, shape your letters such that water flows off them rather than pooling, and hope that, like the Sphinx, nobody in the next several millennia decides to vandalize your masterpiece too badly.

There's something innately appealing—and innately supervillainous— in carving your message into the very crust of the planet itself, and doing it in such a way that it can outlast just about everything else. And if your message is redundantly written in multiple popular languages, that at least gives you some insurance if any individual words are lost or destroyed.

You may be able to improve on the instantiation of this idea. You probably should.

But hold on: we started doing these translations because languages evolve and this increases our chances of one of them still being understood. And while this held true at the hundred- and thousand-year-plus time scales, can we expect it to hold true across tens of thousands of years?

Well, no. Here's where we have to face the inconvenient truth that not a single language in human history has ever survived 10,000 years.

The earliest writing we have is cuneiform (written Sumerian) dating to around 3200 BCE, but that language went extinct around 200 CE, less than 4,000 years later. People were obviously talking about all sorts of stuff before they started writing things down around 3200 BCE, but those early languages are completely lost, and the only evidence we have of them is from reconstructions.

Indo-European languages like English, Russian, Greek, German, Hindi, Punjabi, Armenian, Italian, Latin, Sanskrit, and French (and that's just a partial list) all derive from a mother language we've named Proto-Indo-European, or PIE, which was spoken from around 4500 to 2500 BCE.* It's been extinct for millennia: as speakers of PIE migrated and formed new communities, styles of speaking became accents, accents became dialects, and dialects became new languages. We can partially reconstruct Proto-Indo-European by examining the languages that evolved from it, sort of like how you can squint at frogs and lizards and deduce their common ancestor *probably* had a spine, two eyes, four legs, and laid eggs to reproduce. But without actually having evidence of ancestry, we can only see the broad brushstrokes: the largest features of the language, without the charm, beauty, and grace found in the details. We can find words that sound similar across the family of PIE descendants, from which we might make an educated guess at the sound of the original Proto-Indo-European

---

*Sumerian's not on this list because it didn't evolve from Proto-Indo-European: in fact, we don't know *which* language it evolved from. Languages without a demonstrable relationship to others are called "language isolates," and they either evolved spontaneously on their own (unlikely in most cases), are so distantly related to other languages that we can't find the connection, or they're descended from a parent language that has been lost.

root, and we can see similarities in grammars from which we can try to deduce the rough shape of PIE itself. But these will only ever be guesses, because the original Proto-Indo-European language is gone forever.

And Proto-Indo-European wasn't even the first language! Proto-Afro-Asiatic was probably spoken before it, and although we're not sure, it's possible other languages were being spoken as far back as 50,000 BCE. We'll likely never know for certain, because the speakers of these languages never wrote them down, certainly never carved colossal words into the sides of mountains, and languages evolve so quickly—especially in the absence of writing—that there's no way to credibly reconstruct any traces of these first languages today.

So given all that, given the languages we've already lost and the speed of linguistic evolution and the huge gulf of time you're reckoning with, how can you possibly make yourself understood to people 10,000 years hence?

The U.S. government stands ready to assist you.

Nuclear-generated radioactive waste will still pose a threat to human health in 10,000 years, and if we bury it somewhere, as we have been doing at the Waste Isolation Pilot Plant in Carlsbad, New Mexico, we realized at some point it would be only decent to warn the people who might one day live there not to dig it up. As part of that, the U.S. Department of Energy commissioned various studies and task forces to find a solution to the problem of communicating over such vast distances in time.

One from 1984 recommended the creation of an "atomic priesthood": a group of "self-selective" people who would be entrusted with the knowledge of what's really going on in the area, and who would then be responsible for creating a false trail of superstition designed to keep people away. The message "stay away from this area or vengeful gods will punish you" would be passed down through the centuries with a yearly ritual retelling, by the priesthood, of this new narrative. Those same vengeful gods in the story would also demand that any written signage around the area be replaced and rewritten every 250 years to ensure their language was current.* The only issue—which the report acknowledges—is that it's very hard to get a new religion going, and past attempts at invoking divine threats to protect something—like curses on books and graves, warning anyone away from stealing or disturbing them, respectively—have generally done very little to protect against a motivated looter.

Curses, You Fools!

The idea of protecting nuclear waste with the promise of divine revenge is but another entry in the long and proud tradition of humans using curses as security systems. William Shakespeare's grave reads, *"Good frend for Jesus sake forebeare / To digg the dust encloased heare / Bleste be the man that spares thes stones / And curst be he that moves my bones."* (In 2015, Francis Thackeray, a researcher at the University of Witwatersrand who was interested in studying Shakespeare's remains, suggested he could get around the writer's last wishes by merely examining the *surfaces* of the bones without moving them, before identifying another loophole when he added, "Besides, Shakespeare said nothing about teeth in that epitaph.")

*This same study, "Communication Measures to Bridge Ten Millennia," also suggested that making an area smell nasty wouldn't be a reliable way to keep people away, because future humans are likely to explore the world via robots (or by being safely encased inside robot suits, the language isn't 100% clear) that can't smell. It also—as an aside—mentioned that one day information vital to the survival of the human race, so important that it could affect its very survival, could be added to human DNA via "micro-surgical intervention" . . . but frustratingly, they never indicate what this vital information might be. *Your message, perhaps?*

As for books, in medieval Europe when they were written out by hand and therefore much more labor-intensive and valuable, it was not uncommon to include a curse in the colophon on the last page. These included threats of:

- excommunication ("Whoever takes away this book / May he never on Christ look");
- damnation ("Him that stealeth, or borroweth and returneth not . . . let it change into a serpent in his hand and rend him . . . let him languish in pain, crying aloud for mercy . . . let the flames of Hell consume him forever");
- disease ("Let him also receive by the hands of God the cruelest plague");
- cannibalism ("If anyone take away this book, let him die the death, let him be fried in a pan");
- hanging ("Steal not this book my honest friend / For fear the gallows should be your end, / And when you die the Lord will say / And where's the book you stole away?"); and even
- hanging plus bird torture ("To steal this book, if you should try / It's by the throat that you'll hang high. / And ravens then will gather 'bout / To find your eyes and pull them out. / And when you're screaming 'oh, oh, oh!' / Remember, you deserved this woe").

With the invention of the printing press, books became less valuable, and book curses evolved into bookplates: a simple claim of private ownership that usually, but not necessarily, avoids *explicitly* cursing those who fail to return any borrowed texts.

Two more expert teams were convened, and their combined 1993 report went in the opposite direction, recommending that the truth be told instead, noting that if anyone challenged the dire myth and nothing bad happened immediately—since cancers from the waste could take decades to manifest—then the credibility of the entire project would be questioned. They instead proposed breaking down the message into four levels of complexity, which would help ensure at least some of it is understood 10,000 years hence: rudimentary ("something made by humans is here"), cautionary ("something made by humans is here and it's real dangerous"), basic (the what, where, when, why, and how of the nuclear waste), and complex (highly detailed explanations and records, charts, graphs, maps, diagrams, and so on).

Interestingly, the report suggested that the first two levels of information could be conveyed non-linguistically, through imposing and threatening sculptures and additions to the site: things like giant forbidding monolithic granite blocks, fields of stone spikes coming out of the ground, "menacing earthworks," stones carved to make unsettling ghostly noises when wind passes through them, and so on. Much of the non-linguistic sentiment they thought they could convey through these installations actually doubles pretty well as a supervillain threat: *"This place is a message... and part of a system of messages . . . pay attention to it! Sending this message was important to us. We considered ourselves to be a powerful culture. This message is a warning about danger. The danger is still present, in your time, as it was in ours. The danger is to the body, and it can kill. The danger is unleashed only if you substantially disturb this place physically. This place is best shunned and left uninhabited."*

So: you could always try to communicate a message like this through long-lasting villainous installation art. But let's say you want to do more.

The reports also suggested redundancy: basic information would not just be engraved in multiple languages on granite structures (with room left at the bottom for future generations to add the new ones they speak), but also cast in small ceramics* and buried throughout the site. That way, anyone digging would be likely to encounter them, and they'd increase the chances of the message being found even if the larger objects are degraded or destroyed. This idea of blanketing an area with a particular image or message has been very successfully deployed with branding since the twentieth century. When viewed through this lens, major corporations like McDonald's and Disney are already behaving in a villainous way you'd do well to emulate: your message, after all, needs to survive even after the Golden Arches, Mickey Mouse, and that prying busybody Spider-Man are long gone.

The structure itself is recommended to be in a recognizable geometric shape, so that even if parts of it are missing, the shape of the whole can still be understood: this is inspired by Stonehenge, first built around 2300 BCE. About a third of that henge's stones are missing, but because the rocks were placed in a regular pattern, we can still reconstruct how it originally looked from the partial data that remains.

The most complex information would be stored in the middle of the structure, in a sort of museum also primarily made of granite, where everything that could be explained about the site and the nuclear waste below would be included, with the aid of scale models and periodic tables (which, they hoped, would still be recognizable). The museum itself would be sealed off except by a small stone that could be slid out of the way: large enough for a human to crawl inside, but too small to remove any of the granite instructions through it.

But again, this assumes that the languages could be translated, and the report concedes that complex information cannot be conveyed without it. There's always a chance some future humans would *want* to translate your

---

*The earliest pottery we've found dates back to 18,000 BCE: buried ceramics can last a long time. For added safety, you could make them out of different materials: plastic, titanium, hard glass, etc., thereby hedging your bets that if one material fails, the others might survive. (Titanium isn't cheap, but it does form a patina like bronze does.)

writing—after all, we did when we found hieroglyphics, and mysteries *are* alluring—but there's no guarantee these future people will be successful in their attempts.

So if we can't depend on language, what can we depend on?

Symbols seem to suggest themselves: they're more universal than words, and it seems possible we could build off them to communicate in a sort of ad-hoc language. Plus, we already have the symbol for nuclear radiation—that circle with the three wedges coming off of it—so we're already off to a great start, right? But the problem now is that even *symbols* are interpreted culturally. In 1930s India, the swastika was a symbol of good luck, but in Germany it had a very different meaning. Astronomer Carl Sagan contributed a letter to the project recommending the skull and crossbones as a universal symbol of death and danger, arguing that all humans have them and know what they represent, and their basic shape is unlikely to change. But even here, context matters: the report notes that on a bottle a skull and crossbones means "poison," sure, but on a tall ship it means "pirates," and to medieval alchemists the skull represented the skull of the biblical Adam, and the crossed bones the promised resurrection. A promise of eternal life is the exact opposite sentiment people warning about nuclear waste want to convey! Besides, the nuclear radiation icon was only created in 1946, making it something too new and untested to be staking 10,000 years on. The report even suggests there's a non-zero chance that future people might wonder why we put all this work into bragging about burying a few boat propellers.[*]

---

*In 2007, both the International Atomic Energy Agency and the International Organization for Standardization announced a *new* supplementary symbol for ionizing radiation: this one features the old "propeller blade" icon, but now with waving lines coming off it, which reach out toward a human figure running away from . . . a skull and crossbones.

*A symbol warning of deadly radiation or promising fun self-propelled water
entertainment vehicles, shown beside a figure warning of death,
piracy, or possibly immortal life.*

It's clear that while there are *shapes* that are likely to be recognizable 10,000 years from now, we can't be certain they'll have the same *meaning*. Is the shape of a knife a threat, or is it to indicate cooking is going on nearby? We've experienced this already with cave drawings: we can recognize the human and animal forms in there, but we have to guess why they were drawn and can't even be certain of what they're representing. Here's another example suggested in the report: quick, what's this image trying to communicate?

Those figures could be fighting—or maybe they're dancing. And either way, is it good that they're doing this, or is it a cautionary tale? Are we applauding the figures, or judging them? Is it a promise or a warning, and are we supposed to do what's shown or avoid doing it? With some cultural knowledge—knowing I come from a people who really love to boogie—you could make an informed guess. But we can't be certain people 10,000 years from now will understand our culture, and we certainly can't understand theirs. If we put symbols in sequence to tell a story—in other words, if we use comics—then we still have to figure out a way to ensure those symbols are reliably read in the right order:

Read left to right for a terrifying warning imploring you to not drink water contaminated by nuclear waste, and read right to left for an enticing advertisement about a wonderful nuclear elixir that both calms your stomach and improves your mood!

However, there *is* one set of symbols that is so far timeless for humans: their own facial expressions. Across cultures and time, our babies react with the same faces in response to the same stimuli: happiness provokes a smile, anger a frown, hurt a pained expression, horror an expression of shock, and so on. It stands to reason that a horrified face could still be recognizable as a warning 10,000 years from now—though it could be misinterpreted too, as art or a cathedral of a culture that worshipped fear. But it's a fair guess that a bunch of screaming faces carved in granite will be understood as at least a *possible* threat or warning in the future. And while one of the problems the report struggles with is the fact that any installation warning of danger is inherently *interesting*, and anything interesting will attract curious humans—that's actually great for you! You *want* your message to seem interesting, and you want it to attract attention. That's the whole point of doing something on the scale of Mount Rushmore in the first place!

So by carving your message into a mountain *and* including translations into multiple languages *and* incorporating terrifying installations around the site *and* adding in some scary faces too *and* hiding smaller duplicates of your message in the subsurface at various depths for redundancy *and* ensuring some of them are deep enough that they can't be dug out easily without mechanization *and* repeating your message at various levels of complexity and detail *and* ensuring they're all made out of materials with little intrinsic value that are difficult or impossible to recycle into something more useful (which is to say, granite or ceramics: if you use metals, be sure to hide them deeper so they won't be easily harvested by any post-apocalyptic societies), then you've got the very best chance that this book—and the marshaled knowledge and resources of the U.S. government—can give you to ensure that your message will be received and understood 10,000 years or more into the future.

And even if the people living on our planet 10,000 years from now can't figure your message out, whatever remains of your monument might at least inspire them to worship you as an ancient and powerful god, which is always a pretty decent consolation prize.

TIME PERIOD: AT LEAST 100,000 YEARS

SCHEME: WE'RE BACK TO THROWING IT INTO THE OCEAN, BUT WAY DEEPER THIS TIME

COST: AROUND $33,000 USD PLUS TRAVEL EXPENSES, THOUGH MORE EXPENSIVE IF YOU GO FOR THE BRONZE OPTION TOO

Some say the world will end in fire,
Some say in ice.
From what I've tasted of desire
I hold with those who favor fire.
But if it had to perish twice,
I think I know enough of hate
To say that for destruction ice
Is also great
And would suffice.

　　—Robert Frost (1920)

To ensure your message survives 100,000 years, you must first know what threats Earth is going to face over those 100,000 years. And yes, there is a chance that future generations could turn out *great*, avoiding wars, mitigating famines, curing plagues, and doing all that good stuff that we could do if we really wanted to but haven't gotten around to yet. But thus far nobody has figured out a way to prevent planetary-scale natural disasters, and there are two big ones coming down the pipe that could interfere with your plans.

The first is a new ice age. These happen routinely and cyclically on Earth, and the last time ice sheets were at their maximum—around 24,500 BCE—ice covered most of North America, northern Europe, and northern Asia. The colossal glaciers ice ages produce scour the Earth and remove any trace of human habitation on the surface, so rest in pieces, Mount Rushmore (and also rest in pieces, the Mount Rushmore copy you made in the previous section if it was carved anywhere in the world that faces glaciation). The next glacial maximum is expected to happen sometime in the next 50,000 years, though of course human-generated greenhouse gases could delay that, potentially up to another 50,000 years—right at the cusp of the time period we're considering. So if we want our message to survive 100,000 years or longer, we need to factor in the very real possibility that Earth will be deep into another ice age.

But that's not the only threat. The second one is supervolcanoes—an objectively awesome (and borderline-villainous) name for what is scientifically described as "any volcano so large that it literally maxes out the Volcano Explosivity Index." This maximum score, an 8, is reserved for any volcano that ejects more than 1,000 cubic *kilometers* of material with enough force for it to be injected into the *stratosphere*. That's enough to cover the distance from Toronto, Ontario, to Montreal, Quebec, with a 1km-wide mass of ash, pumice, and lava, and then continuing to cover that area until everything is buried under another 2km of ejecta: enough that if the tallest building in the world, the Burj Khalifa, were there, it would *still* be buried under more than a kilometer of debris. And that's the *minimum* standard for a supervolcano! But of course in real life they affect a much wider area than just a 1km strip between two Canadian cities: supervolcanoes are colossal enough to affect areas as large as North America, and in the process—as we saw in Chapter 4—alter the global climate. These eruptions have never been scientifically observed by humanity (mainly because we've been lucky enough not to be around any that were going off'), but the scars they've left on the face of the planet have been found and studied. And by looking at the history of supervolcanic eruptions—also known as "supereruptions," which is a word I am not making up but am taking from some frankly amazing geology papers—we can see that they happen, on average, once every 100,000 years. Fair to say, it'll be challenging for any message of yours to be found if it's covered by kilometers of ash, pumice, and lava from a future supereruption.

Thus, fire and ice—or, more accurately, supervolcanoes and global glaciation—are the two huge factors limiting where we can put our message. Anything on or near the surface too far away from the equator is out, because glaciers will destroy them. Anywhere too close to tectonic plate boundaries is out, because that's where volcanoes form. This leaves us with tectonically stable zones near the equator, and given that we've already seen how humans love to poke around in everything, you're again going to want to throw your message into the ocean. There, there's at least a chance your message can avoid them for 100,000 years or so.

You could try bronze again, but unfortunately we don't have any evidence that bronze statues can survive this long underwater, patina or not, simply because they haven't existed that long. Human *civilization* hasn't existed that long, and while you could find some humans 100,000 years ago, they'd be deep in the Stone Age. You can still try it if you've got money to burn, but for this range of time you're going to send your message on different material, something far less prestigious than bronze but definitely more abundant, despite the fact that it didn't exist on Earth until we invented it in 1907 CE.

The medium for your noble message shall be *plastic*: fully synthetic plastics, like polyethylene terephthalate.

---

*The most recent supereruption of a supervolcano was the Taupō volcano in New Zealand, around 24,500 BCE. Thankfully, New Zealand was unsettled at that point in time, but nearby Australia had humans on it, who probably noticed *something*.

You've probably heard warnings from environmentalists about plastics in the ocean. The reason they're such a threat is because while synthetic plastics *photo*degrade (breaking up under UV light into smaller and smaller pieces), they don't generally *bio*degrade. For a while we thought there wasn't a single animal on Earth that could digest plastic, given how new the material was, and while we've recently discovered a few—in 2017, a species of wax moth caterpillars were found that could slowly munch their way through plastic—these are exceptional, they're not widely distributed, and they're very, very rare. Unless it's been incinerated, the vast majority of polyethylene we've made still exists somewhere, most of it as waste. It's bad news for the planet, but it's amazing news for a supervillain grappling with the challenge of communicating across deep time! All you need now is someplace that light won't reach for hundreds of thousands of years and which is free from the few plastic-eating beasties on Earth, and it just so happens that (a) light only gets 1km or so into the oceans before it's absorbed, (b) the deepest parts of the oceans are more than 10km deep, and (c) as far as we know and all evidence suggests, nothing that eats plastic can survive down there in the eternal darkness of that high-pressure void.

Perfect.

Some plastic floats—and you can find literal tons of that stuff in the Great Pacific Garbage patch*—but it can also be made dense enough that it both sinks and withstands the pressure at the bottom of the ocean. Dump your polyethylene terephthalate (or "PET") plastic missive somewhere near the equator away from any tectonic boundaries and you've got your best chance your message will *survive* . . . but unfortunately, that doesn't mean you've got a good chance it'll actually be *found*. Remember, the oceans are huge: we've got almost 2.5 times as much ocean floor on this planet as we have dry land. That's a lot of space for a 2m-long piece of

---

*Between 135°W to 155°W and 35°N to 42°N, if you're interested.

plastic to lurk in. To increase the odds of your message one day being noticed, you'll want to put it someplace interesting.

How do you know what'll be interesting 100,000+ years from now? You look at what's *already* interested humans about the planet, and note that one of the most interesting places on Earth—known by both adults and children around the world—is Mount Everest. It's not its geology or history or indigenous species that attracts most of us; it's simply the fact that it's the tallest place there is to climb on this planet: once you've summited Mount Everest, there are no higher mountains left to climb. It's distant and deadly and littered with the corpses of those who tried to climb it and failed, but people try to climb it anyway, because it's there.*

The underwater analog of Mount Everest's peak is the Mariana Trench's bottom, which is the very deepest part of the ocean. Assuming your plastic monument survives its time in the Trench, anything there has a much-better-than-average chance of being found by future sea-exploring humans, because once they have the technology to discover it, they'll be naturally attracted to visit it.

The only catch is, your monument won't be the first bit of eternal plastic waiting for them.

Even though the Mariana Trench is 10.9km below sea level (so deep that if Mount Everest were placed there it wouldn't even break the surface of the water), there's already plastic in it, waiting to be found by future explorers. It wasn't put there intentionally, but ocean probes have already photographed a plastic bag—the kind you might get at the grocery store, use once, and then discard—that's beat you to the finish line.

---

*"Because it's there" is the famous quotation, credited to English mountaineer George Mallory, explaining why he wanted to summit Mount Everest. He and his partner, Andrew Irvine, disappeared while trying to climb the mountain in 1924. Mallory's preserved and frozen body was found in 1999, but Irvine's has never been found. Irvine was carrying an early Kodak camera to photograph the peak, and there's a chance the film could still be developed. Finding it could confirm whether the two of them actually reached the top of the mountain before perishing. So now you've got two reasons to climb Everest: because it's there, and because Irving's fascinating and mysterious frozen corpse is there too.

A message from the past to the distant future. Without language,
this could communicate the idea of "Hah hah hah SORRY??" With language,
on its hidden side, this could also communicate the idea that one could've
once encountered unbeatable savings at Safeway.

In fact, more plastic bags have been found in the Mariana Trench since that one was first discovered in 2018. So yes, you'll be competing with literal trash, but your monument will be bigger, sturdier, and more impressive. And even though the message it contains will be all about you, it'll still be a more inspiring missive to tomorrow than the ones the plastic bags send, which is simply that "we humans haven't been all that thoughtful about our indiscriminate pollution and have somehow managed to get our garbage to even the most inaccessible places on the planet, haha whoops."

The Mariana Trench is tectonically active: it's where the Pacific tectonic plate slides beneath the Mariana plate. These plates move only a few millimeters each year—in respect to the plate it's being subducted under, the Pacific plate is moving just 30 to 57mm a year—which means after 100,000 years, that's still just 3 to 5.7km of movement. Ensure your monument is at least 6km east of the trench, and, even on this time scale, you don't need to worry about it being destroyed by the raging molten rock inside our planet. You do need to worry about it being destroyed by raging molten rock that reaches the ocean floor thanks to volcanism, but it's a calculated risk—and you'll put backups in safer places in the ocean, just in case. And as an added benefit, as that tectonic plate slides ever closer to the Mariana Trench, it also brings your monument closer too, slowly but surely increasing its odds it'll be discovered as we get closer and closer to that 100,000-year mark.

There is one more threat to your scheme, however: *evolution itself.*

The reason plastics don't biodegrade easily is because they're so new. The few things on Earth that can digest them are capable of doing so by accident—their usual food sources just happen to be chemically close enough to this synthetic material—and nothing yet has evolved to exploit the niche of feeding on plastic directly. But just because nothing feeds solely on plastic *now* doesn't mean there won't be something that feeds solely on plastic *eventually.* There's a lot of energy locked up in those polymer chains, and unlike bronze, plastic's made out of organic molecules: carbon, oxygen, nitrogen, and so on, the same elements of life. History shows that when there's a lot of potential food lying around, eventually something evolves to exploit that (like what happened with microbes finally figured out a way to eat wood and bark after millions of years in Chapter 4). And while we can't say for sure, 100,000 years just might be enough time for some bottom-dwelling ocean microbes, living in eternal darkness in a chilly environment where food is scarce, to figure out a way to live off the energy-rich material that increasingly surrounds them in water and litters their ocean floor. Your legacy will be in a race against evolution, and that's pretty badass.

As for cost, you can buy sheets of heavy-duty PET plastic and weld them together to form your object, carving the message directly into the plastic (either by cutting out the letters to make an engraved message, or by cutting out the space *around* the letters to make an embossed one). An industrial sheet of PET over 0.5m tall, 1m long, and 10cm deep costs about $3,500 USD, and you'd need eight of them to get an artifact with dimensions that match your bronze one—though of course the budget-minded supervillain could content themselves with a single sheet of PET, at the risk of it being harder to find at the bottom of the ocean. Assuming you go big, you're looking at $28,000 per monument, plus the cost of welding the sheets together and engraving both sides: we'll budget another $5,000 for that to be safe. At that point, all you need to add on is the cost of taking a transpacific cruise with a sea captain who knows how to keep their damn mouth shut.

An ideal sea captain.

By placing your monuments near the interesting Mariana Trench, and others in safer and more boring spots away from tectonically active areas of our planet, you've got the best chance available to ensure that your hunk of personalized plastic reaches people living here 100,000 years from now.

TIME PERIOD: AT LEAST 1,000,000 YEARS

SCHEME: DITCH THE EARTH ENTIRELY

COST: LESS THAN $40 MILLION USD

> Why are the heavens not filled with light? Why is the universe
> plunged into darkness?
>
> —*Edward Robert Harrison (1987)*

You probably noticed in that last section how the only way we could get close to surviving 100,000 years is to go to someplace that's about as stable, isolated, and far from humans as you can get. Unfortunately, when you're facing potentially *one million years'* worth of meddling, we don't have a lot of tricks left to stay out of humanity's way.

One is fossils: we could try to produce ceramics that roughly match the properties of fossils and bury them, or we could even go so far as to bury metal letters in places likely to *produce* natural fossils: a bend in a sandy riverbed, for instance. We know those can last millions and millions of years. But even fossils don't give us good odds. In fact, the numbers are pretty bleak: 99.9% of all living things die without leaving a fossil, and the odds of anything becoming fossilized are less than one in one billion. But even if you do beat those odds, that's still not enough. To enter the fossil record, your object must be found again, which requires it to be in a place capable of producing fossils, *and* become fossilized, *and* the fossil must make it to the present without being destroyed, *and* it must lie somewhere near the surface where we humans could find it, *and* a human has to dig there, find it, know what they've discovered, *and* share it with the world. It's estimated that only 1 in 10,000 species that have ever lived on Earth have even a single entry in the fossil record. (Other estimates are even worse, putting the odds at 1 in 120,000! The calculation depends on how many species you estimate existed in the past: something difficult to do accurately since *they're not in the fossil record.*)

So, screw Earth. It's too hot there anyway, plus it's also too cold— *somehow at the same time.* You're going to let those foolish mortals have their precious planet while you seek immortality above it: specifically, in a nice stable orbit some 5,900km above it.

Our model here is the LAGEOS-1 satellite, launched in 1976 CE. It's a 400kg, 60cm diameter sphere of aluminum around a brass core, covered with 426 cube-corner reflectors: imagine a golf ball about twice the size of a basketball and you've pretty much got it. It has no electronics or moving parts, as its purpose is simply to reflect laser light back down to Earth. The time between the pulse of laser light being sent to LAGEOS and it being reflected back can be measured precisely, which lets the distance between the satellite and the laser station be calculated exactly, which, when it was first accomplished, unlocked humanity's first measurement of things like continental drift and small irregularities in the Earth's rotation.

And, more useful for our purposes, LAGEOS is in a stable, pole-to-pole orbit, high enough up that it slows only slightly, losing just 1mm of altitude per day. At those rates, NASA calculates the satellite will stay in orbit roughly 8,400,000 years before falling back to Earth. Realizing that this was their chance to communicate with the future, NASA built an information plaque that was carried inside LAGEOS, designed by Carl Sagan. Unfortunately, these two identical pieces of etched stainless steel (18cm tall by 10cm wide) don't say much beyond "this is what the numbers from 1 to 10 look like in binary, and this is maybe what the continents looked like 268 million years ago, this is what they looked like when we launched it, and here are some fun ideas we've been kicking around about what we think they might look like for you, 8.4 million years in the future."*

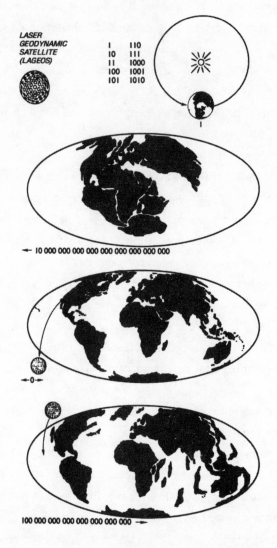

The plaque carried inside LAGEOS, designed to brag to future generations that we knew at least a little bit about continental drift.

---

*Describing LAGEOS's illustration of what the continents might look like in 8 million years as "fun ideas we've been kicking around" isn't unfair: NASA's own 1976 press kit for the launch

This is a *very* achievable way to communicate with the future: it doesn't take special engineering besides the already invented rockets needed for launch, and since we don't care about laser reflection, simply launching *any* mass into a similar medium Earth orbit will do the trick here. It's cheap too: the cost of LAGEOS, including both the satellite and the launch vehicle, was a mere $8.5 million USD in 1976, or around $40 million today—and with corporations beginning to compete for private satellite-launching dollars, prices are coming down. Plus, in rocketry, prices go up with weight, so the lighter you can make your satellite, the cheaper it'll be. Shave off a few of those 400kg from your satellite, and it's money in the bank.* As with all things here, nothing is guaranteed, and there is a chance that as space launches become more common, your satellite may be destroyed, damaged, or harvested ahead of schedule, but then again . . . there's also a chance it won't.

As for what your message should say, we now face the very real possibility that even those basic facial expressions we were relying on in previous sections may no longer be universal. To get an idea of the expanse of time we're trying to communicate over, one million years ago we humans didn't exist, and our protohuman ancestors like *Homo erectus* were just experimenting with things like "fire" and "caring for injured members of their family." We don't know if they expressed emotion in the same way— *maybe?*—which means we definitely don't know if whoever's living on Earth a million years from now will feel things like we do either. We don't even know if they'll consider themselves human! So human facial expressions are probably out, language is definitely out, and we're reduced to

---

describes the future positions of the continents as "little more than guesses," adding, "Our knowledge of [continental drift] should be significantly improved by Lageos." It was! Adorably, this press kit also argues for the inclusion of this plaque by saying, "Whoever is inhabiting Earth in that distant epoch [when LAGEOS returns to Earth] may appreciate a little greeting card from the remote past." They might!

*But not too many: NASA put in that brass core because, though a pure aluminum satellite would've been lighter, it also would've been more vulnerable to the effects of drag. You want your satellite to have enough inertia that it won't slow down too fast in what remains of the atmosphere up there, because that'll bring it back to Earth ahead of schedule.

putting faith in illustrations of the Earth itself and basic counting, as the LAGEOS plaque did. But even so, there's still no guarantee you'll ever be understood.

But honestly, the fact you sent a message a million-plus years into the future is achievement enough that the specifics of what you're saying don't really matter anymore. Go nuts. Write down your manifesto, boast of your greatness, threaten their children, or just show them an engraved picture of yourself reaching the kill screen in 1981's *Donkey Kong*. When your message is found—if it's found—it'll automatically be studied, examined, and made famous. *There are no wrong answers here.*

A tantalizing vision of tomorrow.

TIME PERIOD: AT LEAST 10,000,000 YEARS

SCHEME: TAKE ONE SMALL STEP FOR A HUMAN, BUT ONE GIANT LEAP FOR HUMANITY

COST: ANYWHERE FROM $100 MILLION USD FOR UNCREWED MISSIONS TO $20 BILLION TO $30 BILLION IF YOU WANT TO COME ALONG, GIVE OR TAKE A FEW BILLION

One secret which I alone possessed was the hope to which I had dedicated myself; and the moon gazed on my midnight labors, while, with unrelaxed and breathless eagerness, I pursued nature to her hiding-places.

—*Mary Shelley,* Frankenstein *(1818)*

To send a message 10 million years into the future, you going to put it someplace geologically inactive, free from any sort of destructive weather or even weather at all, and more than 384,000km away from every human on the planet.

You're going to put your message on *the friggin' moon.*

Thanks to its lack of oxygen—or any other atmosphere*—we don't need to worry about rust, so we can use cheaper metals than bronze. There are still environmental factors we need to consider: in a vacuum we want materials that can handle extreme heat and cold, and not expand or contract too much when exposed to either. Given these restrictions, you're going to use the same tried-and-true material used in other spacecraft: stainless steel. It's dense enough to withstand some meteoroid impacts—important for something that's going to be on the moon for a long time—and strong enough to survive the heat-freeze cycle it'll experience for the next several million years: 13.5 Earth days of sunlight where temperatures can reach as high as 127°C, followed by another 13.5 days of darkness, where temperatures can drop to as low as −173°C.†

Our main natural threat here on the moon are the impacts from space already alluded to. Earth has an atmosphere that slows meteoroid and asteroid‡ impacts, burning up all but the largest ones into harmless smoke and fire. On the moon, these hit at full size *and* at full speed. Unfortunately, we don't actually know how often the moon gets hit by space rocks, because we haven't been observing it that carefully for very long. But as a (very!) rough estimate, let's assume that the relative closeness of the Earth and the moon means they face similar threats but proportionally fewer of them hit the moon because it's a smaller target: a quarter of Earth's physi-

---

*This isn't strictly true: the moon has a very *slight* atmosphere—comparable to the atmosphere found up in the orbit of the International Space Station—which is for most purposes functionally a vacuum.

†If you'd like to play it even safer, there are deep craters on the moon, near its South Pole, that are in permanent shadow. While temperatures get cooler there—as low as −247°C—they don't fluctuate as much as they do in direct sunlight, and never get higher than −173°C. However, these craters are also where we believe there may be some long-frozen lunar water ice, which makes them interesting and which makes any monuments there more likely to be disturbed by people in the future ahead of our schedule.

‡Meteoroids and asteroids are basically the same thing: rocky objects from space. The difference is meteoroids are 1m or less in diameter, while an asteroid is anything larger than that, but still smaller than a planet. Now you know! Now your friends can't correct you on this distinction anymore, and you can correct some different friends about it instead!

cal size, and only 1.2% of its mass. Smaller impacts are survivable for your monument, but larger ones are not, and on Earth we can expect an asteroid 1km or so wide—big enough to destroy a large city but not to end life as we know it—to hit every 500,000 years or so. The moon is only about 27% the size of the Earth, and if impacts are scaled proportionally to a cross-sectional area, this suggests they'd happen on the moon every 6.7 million years or so: still well within the time range we're considering here. So, on top of all the smaller impacts, giant asteroids are now also something you need to worry about.

There are two ways to mitigate this threat from above. The first is the ol' "multiple copies hidden around the planet(oid)" scheme, which always increases the odds one of your monuments will survive unscathed enough to be recognized and understood. The second is to bury your monument: on the moon, the lack of geologic activity and liquid water (and any sort of life) means buried things survive a lot better than they do on Earth, and the deeper you bury something, the better protected it'll be against impacts. And while this would usually make your monument harder to find, a buried stainless steel monument is inherently interesting on a planetary body without any others. If it's detected, it could actually ensure your message is found sooner rather than later.

The price of getting your stainless steel monument to the moon comes in at just over $100 million USD: that's how much it cost to send the Israeli Beresheet lander there in 2019. And yes, a gyroscope failure did cause Beresheet to crash into the moon instead of landing softly, but that actually works out to your benefit here: a crash landing *could* help bury your monument. And while $100 million is the steepest price tag we've encountered yet, it's a steal compared to the similar United States' Ranger program of the 1960s, which was a series of probes explicitly designed to crash into the moon (ideally while taking photographs on the way down) and which cost over $1 billion in current-day cash.

But when the whole purpose of going to the moon is to ensure your legacy, it's probably worth doing it in style: visiting the moon and installing your monument personally. In 2019, NASA came up with the price tag of $20 billion to $30 billion USD for returning to the moon by 2024. This number's a little rough: it includes some NASA-specific cost-savings from reusing existing spacecraft that you won't enjoy, but it also includes the cost of a lunar base—something nice to have after a long day communicating with the future, but certainly not mandatory. Still. It'd probably be a super nice place to chill.

The final threat you face is, as always, meddling kids: more precisely, the people alive today and their offspring, down throughout eternity. If anyone *other* than you returns to the moon over the next 10 million years, that means people bounding around, hitting things with golf balls, knocking things over, blowing things up, and generally awrying your best-laid plans. Your advantage here is that lunar travel is still rare: only 12 people have been to the moon, 8 of them are already dead, nobody's been back since December 1972, and the race is still wide open for "first person to travel to the moon in the twenty-first century." This presents an opportunity: the sooner your monument reaches the moon, the greater the claim it has to being an *irreplaceable piece of global human heritage*, some small corner of another world that is forever human.

*A civilization that has, **arguably**, lost its way? (That explosives things is real, by the way. The astronauts blew them up to produce data for their lunar seismometers.)*

Forever Human

This phrase "forever human" has its origin in a speech by William Safire, written for then-president Richard Nixon. It was a speech to be deployed in case the lander failed, leaving Neil Armstrong and Buzz Aldrin stranded on the lunar surface. The speech praises their bravery, speaks of their noble sacrifice in the pursuit of knowledge, and ends with the beautiful sentiment that "Others will follow, and surely find their way home. Man's search will not be denied. But these men were the first, and they will remain the foremost in our hearts. For every human being who looks up at the moon in the nights to come will know that there is some corner of another world that is forever mankind."

It would've been delivered while the astronauts were still trapped but alive, and only afterward would communication be cut from mission control, giving Armstrong and Aldrin their privacy. While we all die alone, Armstrong and Aldrin faced the possibility of dying the *most* alone of any humans in history.

It seems likely that the landing site of Apollo 11—which first brought humans to the moon in 1969 and left behind their footprints, bags of poop, and other artifacts (see sidebar on page 331)—will become a historical site, something revered and preserved by future generations, and possibly even re-created if destroyed.* But part of that is because when Apollo 11 landed, there were just 30 other Earth-made objects on the moon: the remains of crashes both intentional and accidental. Today there are at least 80 different Earth spacecraft scattered across the moon, and when you include the objects from Earth they carried with them—moon buggies, golf balls, a tiny aluminum sculpture of an astronaut, explosively distributed tiny metal Soviet logos,† and possibly a tiny drawing of a penis by Andy Warhol‡—it all adds up to hundreds and hundreds of artifacts.

They can't *all* be special.

As more and more things from Earth end up on the moon, new things there become less and less special, generally in order of precedence. Get as close as you can to the front of the line now, and you help ensure your message will get as far as it can to tomorrow.

---

*On December 31, 2020, the One Small Step to Protect Human Heritage in Space Act became law in the United States: it requires companies working with NASA to not damage any American lunar landing sites. As was pointed out by Michelle Hanlon, an instructor of Air and Space Law at the University of Mississippi, it just happens to be "the first law enacted by any nation that recognizes the existence of human heritage in outer space."

†They were brought to the moon with the Luna 2 mission in 1959, which was the first time any human-made object had touched (and/or crashed into) another world. The spacecraft carried two metal spheres made out of 72 tiny metal shields each, some with the Soviet coat of arms and others with the word "CCCP" on them. They were ejected from the craft and intentionally exploded on impact with the moon's surface, which—assuming they weren't all destroyed—distributed the Soviet branding across the surface of the moon.

‡Hello, and welcome, distinguished reader, to the "tiny drawing of a penis by Andy Warhol" footnote! What happened is an American sculptor, Forrest Myers, collected simple black-and-white sketches from other artists, including Warhol, and then had them printed onto a small silicon wafer, 19mm long by 13mm tall. He claims it was surreptitiously placed beneath the heat shielding of the Apollo 12 landing module by an engineer working on the project, and when he revealed his caper to the world after the crew of Apollo 12 had left both the moon and any tiny lunar art galleries behind them, it was too late to check. The contemporaneous photo of the "Moon Museum" wafer, published in *The New York Times*, has someone's thumb discreetly covering up Warhol's contribution.

Things Left Behind During Apollo 11

When the Apollo 11 astronauts landed on the moon, they collected samples of rocks to bring back to Earth, which obviously increased the weight of their spacecraft. To keep things within limits, they abandoned things they no longer needed, which resulted in over 100 different items being left behind: empty food bags, two pairs of space boots, a hammer, an American flag, and, yes, bags that they'd been pooping into that were now filled with their waste. Nobody wants to take those nasty bags home with them!

Or so we thought. In the time since then, there's been legitimate scientific interest in recovering those poop bags. Because human poop is filled with bacteria, if we were to discover that any of those bags still contain living microbes that somehow survived inside the moon's hostile environment—even in a dormant state—that would give us some idea of how likely future missions are to accidentally contaminate other worlds with Earth microbes.

Incidentally, the American flag famously set up and photographed by the Apollo 11 astronauts was knocked over by the blast of their lander returning to the orbit, and since then it's been bleaching and disintegrating under years of harsh sunlight undiminished by any atmosphere. Some semi-solicited advice: *do not make your message to the future out of a thin strip of nylon.*

TIME PERIOD: AT LEAST 100,000,000 YEARS

SCHEME: YOU'RE GOING TO THE GRAVEYARD OF SPACE

COST: BETWEEN $20,000 AND $365,000,000 USD OR SO, BUT YOU CAN HIT THE LOWER END OF THAT ESTIMATE IF YOU CONVINCE SOMEONE YOUR MANIFESTO HAS ARTISTIC MERIT

> She did it the hard way.
>
> —Bette Davis's epitaph (1989)

Many communication satellites are in a geostationary orbit: a distant orbit, 35,786 km above the equator, where the satellites move in the same direction as the Earth's rotation. From the ground, this has the effect of making the satellites appear to hang motionless in the exact same spot in the sky, which is really useful if you don't want to be constantly adjusting your dish on Earth just to watch some satellite TV.

But there are only so many satellites you can fit into that ring-shaped orbit, so new satellites, when they reach the end of their useful lives, are now generally required by law to save the last of their fuel to leave geo-stationary orbit, go 300 or so kilometers higher up, and then shut down. (It's cheaper to send satellites farther up into space than it is to crash them back to Earth.) This higher orbit, where satellites go to die, is called a "grave-yard orbit": it's unused for any other human purpose, and it's where aban-doned satellites can remain, forever slowly circling the Earth, for hundreds of millions of years or more.

The word "can" there is doing some heavy lifting: the solar system is chaotic, mathematically speaking, and even our best predictions for the orbit of the planets can get unusably inaccurate anywhere between 2 mil-lion and 230 million years out, depending on which planets we're simu-lating and how accurate our initial measurements are. But while we can't expect that *all* satellites, left undisturbed in graveyard orbits, will still be there 100+ million years in the future, we can reasonably assume that *some* of them will be. These satellites may, in fact, be the longest-lasting evidence of humanity's existence in the solar system.

And one or more of those satellites could be *yours*.

Your model here is the EchoStar XVI satellite, a television satellite launched in 2012. In 2027, it's scheduled to have exceeded its useful life-span and be shifted from its geostationary orbit up into a graveyard orbit. What makes EchoStar XVI special is it carries on board a small gold-plated aluminum clamshell case, around 12cm in diameter, inside of which is a silicon disk. Etched onto the surface of this disk are the Last Pictures: 100 different black-and-white pictures, selected by artist Trevor Paglen and his team to represent Earth to the distant future. The case is there to give the disk protection from the harsh environment of space, the gold is there to make the case look like treasure, and the disk is made of silicon for its light weight (always a concern when launching things into space), for its prior success in spacecraft, and for its stability.

At these time scales, you need to worry about diffusion: the way atoms can, over a long-enough time period, move *within* solid substances. When you have two or more different kinds of molecules in a substance, you have the potential for diffusion, and enough movement could, over time, disperse and destroy any information encoded into the material.* To mitigate this, the disk is made of just one material: silicon, and that silicon is constructed so that the entire object is a single unbroken crystal, right to its edges. This single-crystal silicon is both hard and stable, with the added benefit that we've already invented the technology to engrave into silicon at such microscopic scales: we use it all the time here on Earth when making integrated circuits for computers. The result is a wafer that's information-dense, light enough to get *into* space, and tough enough to survive once it's there. And the photos engraved on it are large enough to be discerned with the naked eye, but with much more detail available when viewed under magnification.

Space on the EchoStar XVI satellite was donated, so if you can convince someone with their own satellite company (and a silicon fabrication company) to go along with your scheme, you can actually pull this one off without significant expense. But if you can't, the satellite was launched on a Russian Proton-M heavy-lift launch vehicle (cost per launch: $65 million USD), though it can cost as much as $300 million more to get to geostationary orbit. If your satellite is small enough and light enough—which it might be, since besides your wafer, the only weight it needs to carry is the engines and fuel to move itself from low Earth orbit to the more distant geostationary orbit—multiple satellites can be launched at the same time, which brings down the cost per launch and increases your odds of one of them surviving, like the photos on the EchoStar XVI might, for a hundred million years. And the silicon disks themselves can be manufactured for comparative peanuts: less than $20,000 a pop.†

---

*A piece of art installed in New York City's J. Hood Wright Park is based around this phenomenon. It's 2,000kg of magnesium and aluminum that are expected by its sculptor, Terry Fugate-Wilcox, to fuse together by the year the piece was named after: *3000 A.D.*

†That's the amount that MIT—who manufactured the disk—budgeted for their donation to the Last Pictures project, which includes the cost of materials, lab time, fabrication, a few site

As for what you should put on your disk: remember way back at the mere one-million-years mark, where we saw that at these time scales, it doesn't really matter what you say because the fact that your message *survived* long enough to reach that far future will be the real significance of your object? The same reasoning applies here.

When I asked Trevor about resolving the tension between trying to communicate with the distant future and the knowledge of how hard that will be to successfully pull off, he acknowledged that the photos he helped choose might one day function as an epitaph for humanity, a warning about humanity, or—most likely of all—simply be completely inscrutable to anyone who found them. But that knowledge didn't stop him from choosing them with intentionality. "There are forces at work in this world that are so much bigger than anyone, whether they're economic forces, climactic forces, political forces, and so on, and the possibility of anyone affecting those forces in any way—let alone in a positive way—is basically close to zero," he said. "But that does not give you permission to not care. That does not give you permission to not participate."

And like all supervillains, to the best of your ability, you are going to *participate*.

EchoStar XVI's photos include some subjects you might expect (landscapes, plants and animals, cave paintings, rocket launches, a picture of Earth as seen from the moon), as well as more eclectic choices, such as a screenshot of the first few words of the interactive fiction game *Zork* ("You are standing in an open field west of a white house, with a boarded front door. There is a small mailbox here."), a behind-the-scenes snapshot from 1972's *Conquest of the Planet of the Apes*, an image that Chinese artist Ai Weiwei took of himself flipping off the Eiffel Tower, and a drawing of Captain America taken from 1970's issue #75 of the Marvel Comic *The Avengers*,

---

visits, and a graduate student to help ensure the project proceeded smoothly. Making more than one disk brings your unit costs down!

written by Roy Thomas and penciled by John Buscema, with inks by Tom Palmer.*

And if you want to go darker with your imagery, there's precedent there too: also included on EchoStar XVI are less triumphant images of life on Earth, like a picture of the Ebola virus, of endless rows of chickens tightly caged within the confines of a factory farm, of Japanese children held in American internment camps during World War II, of Vietnamese children with birth defects caused by the widespread wartime use of Agent Orange, and of a CIA Predator drone—photographed from the ground, in Pakistan, as it was flying overhead as part of the U.S. government's remote assassination program.

Each and every one of these images is likely to outlast us all.

A message to the future that is likely to be completely inscrutable, yet also, completely awesome.

---

*It's important to credit your artists, even in space.

Cannibalizing Satellites

There is always the threat of future generations harvesting these satellites for materials or parts. In 2012, DARPA (a division of the U.S. Department of Defense concerned with developing technology for military use; previous successes include remote-controlled cybernetic beetles, the aforementioned Predator drones, as well as the internet) proposed a program called Phoenix, in which a robotic satellite would harvest useful parts (like antennas) from dead satellites in graveyard orbits, and then use them to create new satellites while in orbit. This approach was abandoned in 2015, and the project was given a new focus on inspecting and repairing existing satellites instead.

Nothing has yet been launched, but don't fear: your (functionally useless) satellite should be pretty far down on anyone's To-Harvest-from-Deep-Orbit list.

TIME PERIOD: AT LEAST 1,000,000,000 YEARS

SCHEME: MOVE TO TITAN

COST: $3.9 BILLION USD, GIVE OR TAKE

You think that a wall as solid as the earth separates civilization from barbarism. I tell you the division is a thread, a sheet of glass. A touch here, a push there, and you bring back the reign of Saturn."

—*John Buchan (1916)*

On September 15, 2017 CE, after an almost 20-year-long mission in space, the Cassini spacecraft was intentionally destroyed by crashing it into the planet Saturn in an Emmy Award–winning finale.* But long before that, back in 2005, Cassini's probe Huygens successfully landed on the surface of Saturn's largest moon, Titan. And what's fascinating about Titan is how hospitable it could be to life.

Despite being much farther from the sun—it only receives about 1% as much sunlight as Earth does—we believe it has a subsurface ocean of ammonia-rich water sloshing around under a surface crust of frozen ice. Liquid water occasionally erupts from volcanoes onto the surface, and there are lakes of hydrocarbons on the surface too: the first stable bodies of

---

*An Emmy for Outstanding Original Interactive Program, to be precise. The final flight into that planet was actually called *The Grand Finale*.

surface liquids found on another world. It has a nitrogen-rich atmosphere that's denser than Earth's, so you could (very briefly) walk around without a spacesuit before you suffocated from the lack of oxygen and froze to death in the −179°C atmosphere. It has weather and seasons and a methane cycle that echoes Earth's water cycle, and all this adds up to the remote possibility—just a *possibility*—that conditions might be right on Titan for life to one day evolve, if it hasn't already, somewhere inside that distant moon.* And even if there's no life there now, that doesn't mean life might not evolve there *someday*.

Now we play the waiting game.

---

*Since the Huygens probe wasn't sterilized before landing (the data it sent back indicated Titan was more favorable to life than we'd initially thought, so future missions may have to be more careful), there's an even *more* remote chance that some Earth life could've been accidentally introduced to Titan with the probe's landing and somehow made it into that underground ocean.

Some 5.4 billion years in the future, our sun will enter its red giant phase—during which it'll begin to expand, eventually becoming large enough to engulf Mercury, Venus, and probably Earth, our moon, and any other satellites we had in orbit around those bodies.* When this happens, around 7 billion years in the future, the sun will be putting out enough energy to warm Titan all the way up to around –70°C, which, thanks to all the ammonia there, would be enough for that moon to actually *thaw*. The sun will stay like that for several hundred million years, longer than the amount of time it took for life to evolve here on Earth, which makes Titan the most likely place in the solar system for life to evolve during this era.[†] And if life emerges on Titan, *and* it's intelligent, *and* it can adapt as the sun begins to cool later on, then it could survive long enough to reach the point where it could maybe, just maybe, read something we left behind for it. Of course, the odds of that ever happening are (literally) astronomically low: when you're trying to talk about entities living over a billion years in the future, nothing is promised except the overwhelming likelihood of failure. But there's still a chance.

---

*At this point I must apologize for wasting your time with all those previous talk-to-the-future schemes contained in this section—all of which would now be both engulfed and destroyed by our red giant sun—especially if you've been somehow implementing them live as you read through this book.

[†]One of the papers exploring this possibility ("Titan under a Red Giant Sun: A New Kind of 'Habitable' Moon," by Ralph D. Lorenz, Jonathan I. Lunine, and Christopher P. McKay) cannot resist calling this colder version of Earth's primordial soup a "primordial gazpacho," and *I* cannot resist telling you about that.

The Last Death on Earth

The good news is that nobody on Earth will die when the sun becomes a red giant 5 billion years from now! The bad news is that's because everyone on Earth will already be long dead, having perished a mere 1 billion years from now when the sun warms up enough to boil the oceans and ensure that there's no place left that's even marginally habitable for humanity. And at most 1.8 billion years after that, even small colonies of single-celled life-forms that managed to hang on in isolated and sheltered areas (think cool mountaintops or icy caves) will also die. The water there will eventually evaporate too, and we don't know of any form of life that can survive indefinitely without it.

That final moment of death, 2.8 billion years from now, is when the story of life on Earth comes full circle. Having once boasted life-forms reaching the furthest evolutionary heights of dogs and dinosaurs and weird bugs and humanity, the last survivor of Earth is likely to be a simple bacterium: invisible to the naked eye, single-celled, alone, and resembling nothing more than that very first microorganism that evolved here some 6.8 billion years before . . . until it too passes from this world.

And you're going to take it. You're going to pull off something that the combined might of NASA, the European Space Agency, and the Italian Space Agency couldn't do: you're going to send a message to the icy moon of

Titan, across a billion kilometers of space and a billion years of time, so that you might communicate not with humans,* but with *intelligent life that hasn't even evolved there yet.*

That Cassini-Huygens probe that launched from Earth in 1997 didn't contain any messages for alien life, but that wasn't always the plan. As early as 1994, Jon Lomberg, the design director for NASA's Voyager Record (we'll get to that shortly) was leading a very ambitious scheme to contact any future life that might one day evolve on Titan. The plan was to inscribe information onto a circle-shaped wafer of industrial single-crystal diamond: 1mm thick, 2.8cm in diameter, and 4.32 grams in weight. Scientists from both the Jet Propulsion Laboratory in America and the National Research Council in Canada collaborated on figuring out a way to etch a microscopic message into diamond, which hadn't been done before, and they succeeded in demonstrating it was possible. Diamond is obviously an expensive material, but it was chosen for several reasons: it's the hardest material we have, so it'll resist abrasion and weathering better than anything else we know of,

---

*All human life on Earth already went extinct in an earlier sidebar. Sorry you didn't notice it. Sorry!

plus it's inert and doesn't react with any known materials on Titan. And best of all, it's transparent, which means a second layer of diamond could be put on top, protecting the message from the ages while keeping it legible to anyone who found it. It's not guaranteed, but there's a non-zero chance it could last the amount of time we need here.

At this scale there was only enough room for a few images, and after you placed some of the place-and-timekeeping ones—images of the stars and of the rings of Saturn (both of which slowly change over time) to help date what era it came from, a map of the solar system, a schematic of Earth showing the current position of the continents, and so on—there was really only room for one or two more images, conceived (and titled) as a Portrait of Humanity. This image would be shot stereo-optically—two cameras taking a photo at the same time—both for redundancy, and for the fact that if both images survived, it would give the Titans the chance to notice the slight differences in the two images and deduce they should be viewed together to produce a 3-D image, like those found in View-Master™ toys.* This also had the scientific benefit of indicating the scale of items in the background!

The photo was carefully composed to show a group of humans of various ages, genders, and ethnicities, all on a beach in Hawaii, with one of them holding a replica of the diamond wafer so that everyone's sizes relative to that reference item would be clear. There would be a nursing mother to show how humans care for their young, and a set of mixed-gender twins to help indicate sexual differences between humans. In December 1996, over a thousand stereo-optic candidate photos were taken by photographer Simon Bell in Hawaii, the winner was chosen by NASA, and all that was left was to engrave the diamond wafer and attach it to the spacecraft so that it would be ready for the October 1997 launch.

---

*Of course, there's nothing to say the Titans will have developed vision, binocular vision, or even View-Master™ technology, but it at least gives them the chance to recover *some* depth information.

Six copies of the wafer were needed: one for Huygens to land with, one for Cassini to have on board (at that point the decision to destroy it in Saturn hadn't yet been made), two more for each of the backup spacecraft, plus two for testing. The costs of these six wafers were estimated to be around $60,000 USD in 1995, or around $100,000 today—a pittance on top of the $3.9 billion the mission would finally cost.

And that's about when everything fell apart.

Funding for the diamond wafer project was provided by the Fuji Xerox corporation, who in return for their money expected to have their logo engraved on the diamond too. Had NASA agreed to this—and they initially did—it could've made *this* uninspired, busy, and too-wordy affair the longest-lasting logo, and perhaps longest-lasting evidence of human existence, in our solar system:

THE DOCUMENT COMPANY
FUJI XEROX

*This corporate logo came closer than most to representing,
for eternity, all of humanity's struggle and triumph.**

However, when NASA decided not to allow the logo to be included, funding was put in jeopardy. On top of that, there were the usual human disputes of who had final say over the image, petty political maneuverings by NASA insiders over who'd get credit for the project, and so on. Eventually it was simpler and cheaper to just *not* send an etched diamond wafer into space so it could wait around on Titan for 7 billion years, so that's what happened.

But you don't work for NASA, *and* you're a supervillain! You're unbound from the petty scruples and ethics that bind other humans, which means you're more than willing to let the good folks at Fuji Xerox Company think they're gonna have their logo on Titan for billions of years . . . right up until the moment your spacecraft launches from Earth with a diamond wafer engraved with *your* logo and *your* face and *your* message for whoever inherits our solar system next, and without even the merest *hint* as to who those future Titans might one day employ to help meet their document-duplication needs.

---

*This logo was abandoned in 2008 in favor of a new one that drops the slogan, unifies the typeface, and, because it was 2008, adds in a swooshy ball.

TIME PERIOD: AT LEAST 10,000,000,000 YEARS

SCHEME: DITCH THE ENTIRE SOLAR SYSTEM AND TRAVEL INTO THE UNKNOWN

COST: $5 BILLION USD, GIVE OR TAKE

This hour will pass—all passes,
On this life's fleeting scene;
But still the future glasses
All that the past has been.
— *Letitia Elizabeth Landon (1835)*

You are trying to send a message *10 billion years* into the future. That is *wild* . . . but it's not necessarily *impossible*. We began this effort at preservation by uploading your information to the internet, where everyone can see it, but have since moved it to Earth's orbit, and then to its moon, and then finally to hiding it on the moons of other worlds over 1,300,000km away, hoping it'll be left alone long enough for intelligent life to evolve and figure it out. With every step toward a deeper time, we've moved farther and farther away from everyone else on Earth. And when you're staring down 10 billion years, even staying in the outreaches of our solar system doesn't make sense anymore.

By this time, Mercury, Venus, and Earth are all likely gone, and some of them could've smashed into the others on their way out: *thanks for nothing, Mercury.* The sun—once a red giant—has blown its outer layers off into deep space and collapsed into a dense and cooling white dwarf. *The solar system sucks now.* It's so unpredictable that we don't even know how many planets it'll have, and unpredictability wreaks havoc on the eternally stable orbits we desire.

Your solution is to leave the entire solar system behind, moving out into deep space: an area of the universe famous for being mostly empty. You don't have a particular destination. You don't have a particular hope of being found. And even if you are found, you certainly can't expect that whatever's out there in the depths of the universe 10 billion years hence will understand you. *But that's precisely how you win.* Because when you follow the trail blazed by the Voyager spacecraft and send something from Earth into deep space, you won't just be sending a golden record or some other monument to your greatness (and maybe humanity's greatness too) (you know, if there's room).

You'll also be sending your beautiful corpse, chilled to just above absolute zero and preserved from decay in the frozen depths of space.

Many bacteria can't live without oxygen, and even those that can are certainly not active in temperatures so close to *literally the coldest it's possible for anything to be.* Your dead body will remain very similar to how it was in life, at least at the macro layer, for a very, very long time.* This is the cryonic suspension of Chapter 8 taken to its universe-scale conclusion, and without the need for constant coolant and attention. And not only will your space corpse provide a wealth of information about life on Earth to anyone lucky enough to find it, as it tears through the unfathomable

---

*Your body *is* made of different materials, however, which means you do need to worry about atomic diffusion and your different atoms moving around over long periods of time. But your advantage here is if atoms were cooled to absolute zero, they wouldn't move at all, and unlike spacecraft in orbit near suns, your body will be just a *shade* above that temperature.

depths of outer space, it'll also become—if everything goes as planned—the one thing in creation that serves as the final message, eternal epitaph, and consummate representation of and/or monument to *all life that has ever or will ever exist on Planet Earth*.

You're a supervillain, so it's a pretty fair bet you've said the words "Fools! I'll show you! I'll show you all!!" at some point in the past. But are you truly committed to doing that? Because becoming one of the last extant things this planet ever produced is the *ultimate* way of showing them all.

Here's how you pull it off.

The models for your scheme are the two Voyager spacecraft. The first, Voyager 1, launched in 1977 on a mission of solar system observation. At the time of this book's publication, it was over 22 billion kilometers from Earth, and moving so fast (over 60,000km/h) that it's gaining more than half a billion kilometers more each year. It's both the farthest and one of the fastest objects ever made by humans, it's the first object we ever managed to cross into interstellar space, and it's still active. (But not for long: by 2025, the radioactive material powering Voyager 1 will likely have decayed so much that it can no longer produce enough electricity to run even one of the spacecraft's instruments, and the spacecraft will finally shut down entirely.) But even after Voyager 1 powers off, it'll continue to rip through the universe at its speed of over 16 kilometers *per second*, and since it's not aimed at anything in particular, and since space is so empty, there's a chance it could keep doing that just about indefinitely.

Knowing it would end up tearing through deep space, NASA famously included on both Voyager spacecraft a gold-plated copper record, 30cm in diameter, held in a protective aluminum jacket. On the jacket were inscribed instructions on how to decode this data, assuming whoever finds it has a similar mathematical and sensory system to us and can understand that it's meant to be played at 16⅔ revolutions per minute. On that record itself were encoded 116 photographs and 90 minutes of music from around the world, greetings in 55 different languages, and other sounds from Earth.*

What Else Is on the Voyager Records?

Less famously, the Voyager records also include an hour-long recording of the electrical activity in the brain of the project's creative director Ann Druyan, recorded by electroencephalographic monitoring, converted into sound, and compressed down to one minute of audio—done under the incredibly optimistic reasoning that nobody could say for certain that an alien a billion years from now *wouldn't* be able to understand them. During the time the electrical activity in her brain was recorded, Druyan thought about ". . . the history of ideas and human social organization. I thought about the predicament that our civilization finds itself in and about the violence and poverty that make this planet a hell for so many of its inhabitants. Toward the end I permitted myself a personal statement of what it was like to fall in love." There's already been some pretty out-there information sent into the universe on physical media, so don't let anyone tell you that the

*Jon Lomberg (also of the Titan diamond wafer scheme) told me it was a miracle the Voyager records happened at all, given that they had *six weeks* on the project from start to finish. But it was also a blessing: that tight time frame left NASA bureaucrats with a take-it-or-leave-it choice that prevented them from meddling too much in its content. NASA decided to take it, and the work of Jon and the rest of the Voyager record team will likely become one of the longest-lived pieces of human art ever produced. He contributed to the long-term nuclear waste messaging project too!

information you want to send isn't valid, up to and including recordings of *your* brain waves as you remember the greatest world domination hits from your life.

Just don't bet on these thoughts ever being recovered: the data's simply not there. Electroencephalographic machines (or EEGs) simply measure electrical activity found in different areas of the head, and at a relatively coarse level. They can notice the areas of the brain that activate when you, say, raise your right hand, but EEGs are nowhere near the resolution required to reconstruct actual thoughts. It's roughly analogous to studying an orbital image of China in order to find out if your friend who was living there at the time was feelin' sad that day.

Here's where things get a bit dicey: the records on both Voyagers were projected by NASA engineers to last one billion years (for the side facing space and the dust and micrometeor impacts it offers), with the interior side lasting potentially much longer.* If both sides fail, there's still a patch of ultrapure uranium-238, 2cm in diameter, electroplated onto the record's cover as a timekeeping fail-safe. Even in the absence of a fully functional record, as long as some of that patch survives—and as long as aliens have figured out that the half-life of uranium-238 is 4.51 billion years—they'll be able to date the age of the spacecraft for several dozen

---

*Recent research by Nickolas O. Oberg and Sebastian Roelenga (shared with me in a preprint article by Oberg) estimates that the internal side of the Voyager 2 record could survive up to the merger of the Milky Way and M31 galaxies, some 5 billion years in the future. After that point, it's hard to credibly simulate the course of the spacecraft, though if it's flung out of our galaxy and into a region of space that's essentially dust-free and avoids passing through any planetary systems or encounters with interstellar objects, the interior side of the record might last a hundred thousand trillion years or more: that's a one followed by 17 zeroes.

billion years at least, and we're at least communicating *something*. But even so, we're still running up against the fact that we don't actually know what happens to most materials over such a huge expanse of time: we're just making educated guesses, because we haven't been around long enough to watch it happen. It could last longer, or it could last for less. The Earth itself is only 4.54 billion years old, which means we're reaching for lengths of time more than twice the age of our planet! And of course, there's always the threat of other materials too: while space is *mostly* empty, there's still interstellar dust and radiation and other passing bodies we've mentioned that could, over time, damage anything to the point where it's no longer recognizable. It's a risk. But if you want to talk to someone alive 10 billion years from now, *you're going to have to take a few risks.*

Your plan is to first produce your own custom, likely self-aggrandizing version of the Voyager record (which cost $18,000 USD in 1977, or around $80,000 today), filled with the glory of your life, some of your most flattering portraits, and perhaps a few of your favorite bops. Then, you'll finance your own spaceship and launch vehicle (again, similar to what was built for Voyager), having it built, launched, and delivered into deep space out past the edge of Neptune. The cost for both Voyagers was around $865 million in 1977, or about $3.8 billion today. Since you only need one Voyager, this is a pretty generous estimate, but we'll still add on another pretty arbitrary billion and change—to cover the modifications necessary to fit you inside the spacecraft, and to gussy up your Voyager in style. Remember: you're building your own tomb, so money should really not be an object here. After all, you can't take it with you.*

Thankfully, altering the Voyager spacecraft to fit you inside shouldn't be too difficult. The spacecraft itself—without the various booms and

---

*You can take *some* of it with you, of course, and yes: the bit you're bringing with you to space is being taken in a more permanent way than any other human has managed to do in all of history, but you still can't take it *all* with you. Even a supervillain's spacecraft has weight and size restrictions.

antennas—is about the size and weight of a subcompact car, which gives plenty of room to squeeze you in, especially if you remove some of the cameras and magnetometers you won't need on your voyage. Beyond that, there are very few alterations necessary, because unlike most human-crewed spaceflight, *we don't actually want to keep you alive.* To the contrary, our goal is for you to get into space and then die as quickly as possible, ideally with the record strapped to your body, held in your hands, and positioned in such a way that your eternal corpse looks both impressive *and* attractive.

While the record is estimated to last a billion years, we don't exactly know how long human bodies can last in space, beyond "possibly indefinitely": space is big, empty, anaerobic, and very, very cold, and those are the ideal properties for any medium used to store human remains. But on the other hand, we also know that, thanks to the second law of thermodynamics, even when stored close to absolute zero in empty space, your body isn't going to get any *more* organized. And that means it could eventually fall apart. There may very well still be some of you—perhaps a large chunk of you—intact after 10 billion years, but there's no way to say for certain.

And as for what happens next?

Well, you'll either get torn apart as you fall into some distant black hole, or you won't. You'll either burn up in another sun's corona, or you won't. Some parts of your body will either miraculously land in the primordial ooze of some distant planet and forever change the sequence of events on that world, or they won't. And as the universe moves toward its final and ultimate conclusion, you will either be found, or you won't. Your only consolation is that these risks also apply to Voyager 1, Voyager 2, and any other wandering interstellar spacecraft.

Either way, your investment of $5 billion has created a tomb that will likely never be plundered, and made for you a new kind of immortality— one unlike anything available on that lost and distant world that, in some forgotten past, once gave you life.

R.I.P. EARTH, 4.5 BILLION BCE – 7 BILLION CE (ISH)

A PRETTY GOOD PLACE TO STORE A BUNCH OF STUFF

Voyager, Revisited

Jon Lomberg believes that the Voyager record is a piece of art and that as much as it was a message *from* Earth, it was a message *to* Earth: a reflection of how we see ourselves, or at least how we wished to see ourselves. "There's a victorious element to it: whether it's found or not, and most likely it won't be, but it almost doesn't matter," he told me. As to its content, he said, "We had a much more optimistic vision of the future in 1977; we had a much more optimistic and positive image of extraterrestrials in 1977. When we had the opportunity to do a Voyager record again on the New Horizons spacecraft, NASA shied away from it . . . one of the reasons was there's now been a growing sentiment that the universe may not be a friendly place. We may not want to announce ourselves to the universe . . . Nothing has changed in the universe, and we haven't discovered anything that leads us to believe one way or the other, but our *culture* has changed, and *we* have changed."

One of the first decisions the advisory board Jon put together for the Voyager record was that they would show humanity on a good day. "There's nothing in the Voyager record that indicates or suggests the host of problems that we face," Jon said. "And that was a decision that was taken virtually unanimously. The analog was, you go on a dating site, you don't list all your flaws. When you meet somebody, you don't tell them all the terrible things you've done—especially if it's in an obituary. An obituary rarely talks about somebody's worst days, and this was our obituary." But when Jon was doing early work on a Voyager Record 2.0 for the New Horizons spacecraft, a different decision was made: "When I asked the international group of advisers—very multidisciplinary— whether we should follow [Voyager's "Earth on a good day" example], they were unanimous in saying no. A picture of Earth now that didn't include the problems we face in climate change and everything else would be a dishonest picture." Viewed through this lens, a message from Earth *you're* sending—your dead body clutching some gold in the empty depths of space—could actually be one of the more honest self-portraits we've ever sent into the universe.

Incidentally, NASA hasn't gotten out of the long-term message business, but these days messages included on spacecraft tend to be much less ambitious. They invited anyone online to submit their name to be microscopically engraved on two silicon chips glued to 2018's InSight Mars Lander, and 2,429,807 people took them up on that offer. Of course, it's hard to imagine any alien life coming across that message and thinking too highly of a species that, when encountering a chance to speak to eternity, decided the most important thing to communicate were the names of a couple million internet users, *especially* if they didn't even throw in a dead body to go with.

TIME PERIOD: 100,000,000,000 YEARS AND ABOVE

SCHEME: HOPE FOR THE BEST

COST: FREE (ASSUMING YOU'VE ACCOMPLISHED THE PREVIOUS SEND-YOUR-CORPSE-INTO-DEEP-SPACE SCHEME)

It's the end of the world every day, for someone. Time rises and rises, and when it reaches the level of your eyes you drown.

—*Margaret Atwood (2000)*

Even with our best efforts, and even with your frozen corpse hurtling through the depths of interstellar space, we can't hide from the ravages of time forever. When you peer a *hundred billion years* into the future, what you see is the beginning of the end of the universe.

So here's what that looks like.

While we don't yet know *exactly* how our universe will end, one of the leading theories is that it will simply continue to expand and cool indefinitely, a scenario known as the heat death of the universe. That goes down like this:

- In 100 billion years, the universe's expansion—already recently discovered to be accelerating faster than we thought—will be such that, except for the stars in our local supercluster of galaxies, no other stars will be visible in the sky—even with telescopes.* This naturally limits the pool of alien life-forms that can find your dead and frozen corpse.

- In 336 billion years, any individual star outside a galaxy will become isolated in what appears to be a dark, empty, and indifferent universe: someplace 200 million times larger than the one we know, but with the same amount of matter and energy to go around inside it. This further limits the pool of alien life-forms who can find your dead and frozen corpse.

- In around 1,060 billion years, the universe will be so large—600 septillion times larger than it is now[†]—and with an average particle density so low that individual particles in intergalactic space will be effectively isolated, never again to interact with any other matter in the universe. This could include your dead and frozen corpse—or whatever particles remain of it—as it careens through this void.

- Somewhere around 100,000 billion years in the future, star formation will stop, because there will be no collections of gases large enough left to form them. The universe is now inconceivably large: compared to today, it's a one followed by *2,554 zeroes* times larger. All that the remaining stars can do is use up their fuel and die as the universe slowly grows dark. When that happens, what remains of your dead and frozen corpse will become even

---

*While nothing in the universe, to our knowledge, can go faster than light, that doesn't mean the universe *itself* has a speed limit on its expansion. Imagine two ants on the surface of a deflated balloon. As the balloon expands, the ants get moved farther and farther apart, despite not moving themselves. And if this balloon accelerates faster than the speed of light, the two ants will eventually disappear from each other, because light from one ant emitted *after* the universe reached faster-than-light expansion will no longer reach the other: the two ants are simply moving apart too quickly for even light to catch up.

[†]A septillion is far beyond human concepts of scale—but to get an idea of how much bigger this is, there are a trillion trillions in one septillion: it's a one followed by *24* zeroes.

more frozen, and, in the new all-encompassing darkness of space, even harder to spot.

- If protons themselves can decay (and they might, we don't know if they can't, or if they're all just too young to have started decaying yet), then the very building blocks of the atoms themselves might start breaking down into lighter subatomic particles around 10,000,000,000,000,000,000,000,000 billion years in the future. Unfortunately, this will have the effect of destroying what's left of your dead and frozen corpse.

- Assuming protons decay, then black holes will dominate the universe as soon as 10,000,000,000,000,000,000,000,000,000,000 billion years in the future, which will, again, destroy anything that remains of your dead and frozen corpse. If protons don't decay, the universe will still end up dominated by black holes, but it'll take significantly longer.

- We believe that black holes actually *evaporate* over time, through a process called Hawking Radiation.* And as soon as 10,000,000,000,000,000,000,000, 000,000,000,000,000,000,000,000,000,000,000,000,000,000,000, 000,000,000,000,000,000,000 billion years in the future, all those black holes will evaporate entirely, leaving the universe a place that's mostly empty, save for a few isolated particles.

- At this point it's quite difficult to find any good news about what remains of your dead and frozen corpse.

---

*The idea is that sometimes, subatomic particles may spontaneously manifest in the universe: one particle of matter and one of antimatter. Normally these two particles would mutually annihilate a moment later, but near black holes, one could be drawn into the hole while the other escapes into space. This would cause black holes to radiate particles over time, which—in order to keep the first law of thermodynamics happy, which states that energy cannot be created or destroyed—would also result in the black hole getting ever-so-slightly smaller with each particle emitted.

But despite this upcoming death of all life and destruction of everything anyone has ever known, it remains inarguable that the best, most impressive, most memorable, *and* most cost-effective way to communicate with the future is to die in a properly constructed spacecraft while that spacecraft is being blasted into the inconceivable depths of space, thereby giving you the chance that at least some of your corpse hangs around long enough to see the beginnings of the heat death of the universe. That investment-to-immortality ratio *alone* is by far the best in this book.

Good luck!

Conclusion

YOU ARE BECOME SUPERVILLAIN, THE SAVIOR OF WORLDS

I am so used to plunging into the unknown that any other surroundings and form of existence strike me as exotic and unsuitable for human beings.

—*Werner Herzog (1981)*

This book has taken you from "curious reader of funny nonfiction books" to "someone who now knows how to mess with the weather, take over a country, become immortal, and ensure that they're one of the last things existing in this universe, *all while piloting a floating geodesic sphere from the comfort of your dinosaur steed.*"

Not bad, if I do say so myself.

In the days that follow, for the rest of your now-supervillainous life, I ask that you remember the lesson of these pages, which is simply this: while the world may be large and complex and hard and unfair, it's also *knowable.* It can be understood—as a species, we've spent thousands of years working on doing just that—and once it's understood, it can be directed, it can be controlled, and it can be improved. Aspirations that once seemed wild and impossible quickly become routine once accomplished—and it's through that very process of slipping into mundanity that mad science becomes simply ... *science.* Boring, regular-old science, just another part of our world in which we fly through the air, chat to people on the other side of the Earth, send robots to Mars, and come up with multiple vaccines for a global pandemic within the space of a year.

The truth is, as complicated as our planet now is, there's *still* room for individuals, acting alone, to change the course of history.* We all started out as stupid babies who couldn't survive without constant care, but within the space of only a few years, we learned to walk, how to not poop our pants, and how to understand and speak a language from scratch *simply by observing it being spoken around ourselves.* In other words, we begin our lives as curious, precocious, ambitious autodidacts, and some of us simply never stop, continuing to learn, making ourselves just a little bit more prepared for that one fine day—if and when it comes to that—when we find ourselves holding *the very fate of humanity itself* in our hands.

You, for example, have just finished doing exactly that.

So go ahead and dream big. Chip away at the impossible until a tiny piece of possibility breaks off. You have it within yourself to build that better world, and like all supervillains, *you aren't going to ask someone for permission first.* I believe in you. Better: I believe in *us.* It's like what Lex Luthor once said to Superman:

"Trust me, Clark. Our friendship is going to be the stuff of legend."

---

*Though it's worth pointing out the examples I just gave (flying through the air, chatting with people on the other side of the Earth, sending robots to Mars, and inventing vaccines) were all *group* efforts. That's right: your supervillain guide is introducing a twist in its last pages, setting up a sequel in which you team up with others to take over even more planets than any one villain could alone. Can't wait.

Go get 'em, tiger.

THANKS AND ACKNOWLEDGMENTS

The one duty we owe to history is to rewrite it.

—*Oscar Wilde (1891)*

Usually at this point authors apologize for any errors in their books, assuring their readers that while many people have helped with the manuscript, any mistakes that remain are their own and were introduced by accident. Well, not me. While many people helped me bring this book to publication too, any errors that remain are on purpose actually, as part of a secret villainous scheme to keep all the heists to myself.

But I do still have some people to thank!

First, I want to thank my editor, Courtney Young, for her amazing feedback at every stage of this book, from concept to execution. Your favorite author is nothing without their editor—*nothing!*—and Courtney is head and shoulders above the rest. You're the best, Courtney.

Thanks of course to my long-time friend and new-time coworker Carly Monardo, whose illustrations brought this book to life. The supervillain character was her design, and she instantly became the heart of the book.

Thanks to the professionals who spoke with me and answered my very suspicious-sounding questions during the writing of this book, including Trevor Paglen, Dr. Blake Richards, Simon Bell, Nick Oberg, and Jon Lomberg. (I promised Jon a quick set of questions and we ended up talking for three times longer than expected before I felt guilty enough about monopolizing his time to end it: he's a funny, thoughtful, and fascinating man, and the work he and his team produced for the two Voyager spacecraft is something that still strikes the imagination today.)

Huge thanks to Christopher Night, who fact-checked this entire book and caught some mistakes that I definitely put there on purpose as a test and *not* because I'm bad at books, and to whom I owe an apology to for writing a book that's so difficult to fact-check. *The real supervillain was within me all along.* Thank you so much, Christopher!

Emily Horne has always been my first reader, the voice of well-informed, educated, worldly reason, and she's been unfailingly great once again at pointing out the parts of this book that worked and the parts that didn't. Thank you, Emily: this book is better because of you. And thanks to all the other friends kind enough to offer to read and give me feedback on earlier drafts of this book, thereby also improving it and making me appear more competent: my wife, Jennifer Klug; my father, Randall; my brother, Victor; and my talented friends, including Janelle Shane; Chip Zdarsky; Mike Tucker (whose name I misspelled when thanking him in my last book, sorry Moke); Jon Sung; Mike Todasco; Marguerite Bennett; Dr. Priya Raju; Dr. Katie Mack; Randall Munroe; Ryan Junk; Jon Manning and the others at and adjacent to Secret Lab; and Gretchen McCulloch and the others in the Scintillation Regular Reading Group. Particular thanks are due to Kelly and Zach Weinersmith, who shared some of their research notes with me when they saw areas of overlap with their own work. (They're working on a book that should be coming out a year after this one, so hey, now you've got a future nonfiction book lined up too!) Zach's feedback on an early draft was particularly helpful—the man is an unstoppable machine of good ideas and suggestions. Honestly, my sincere advice is to simply make friends with people who are smarter and more talented than you; it's been working out great for me so far.

Thanks to S. Qiouyi Lu and Arpan Malviya for their translations in Chapter 9 into Mandarin Chinese and Hindi, respectively. Thanks to Elijah Edmunds for his help in identifying the artists who created the Avengers panel sent into space. And thanks to Nikki Rice Malki and David Malki, who helped me through one of the problems encountered in this book. No supervillain is a superisland.

And finally, sincere and heartfelt thanks to my literary agent and fellow author Seth Fishman, whom I completely neglected to thank in my last book, a fact he discovered when pulling a copy of *How to Invent Everything* off a bookstore shelf and flipping to the acknowledgments in order to impress a prospective client with how grateful all his other clients were for him and his services. Whoops. I therefore owe him double-thanks for this book.

Thanks again to my literary agent and fellow author Seth Fishman.

And thanks to you for reading this book! See? Even though the book is over, you kept going through this extra bit at the end and got a personal thanks out of it. Reading pays off again!

Ryan North
Toronto, Canada
2022 CE

BIBLIOGRAPHY

Finally, from so little sleeping and so much reading, his brain dried up and he went completely out of his mind.

—*Miguel de Cervantes,* Don Quixote *(1605)*

The following is a selected bibliography: accessible sources I thought a supervillain might find particularly interesting regarding the subjects discussed in each chapter. The full bibliography, too large to fit in this book, is available online—assuming you have not yet destroyed the internet—at www.supervillainbook.com.

CHAPTER 1: EVERY SUPERVILLAIN NEEDS A SECRET BASE

Craib, Raymond. Unpublished book notes for agrarian studies readers. Cornell University Department of History. n.d. Accessed May 2021. https://agrarianstudies.macmillan.yale.edu/sites/default/files/files/CraibAgrarianStudies.pdf.

Doherty, Brian. "First Seastead in International Waters Now Occupied, Thanks to Bitcoin Wealth." *Reason.* March 1, 2019. https://reason.com/2019/03/01/first-seastead-in-international-waters-n/.

———. "How Two Seasteaders Wound Up Marked for Death." *Reason.* November 2019. https://reason.com/2019/10/14/how-two-seasteaders-wound-up-marked-for-death.

Ells, Steve. "Endurance Test, Circa 1958: 150,000 Miles Without Landing in a Cessna 172." Aircraft Owners and Pilots Association. March 5, 2008. https://www.aopa.org/news-and-media/all-news/2008/march/pilot/endurance-test-circa-1958.

Etzler, John Adolphus. *The Paradise Within the Reach of All Men, Without Labor, By Powers of Nature and Machinery: An Address to All Intelligent Men, In Two Parts.* London: John Brooks, 1836.

Foer, Joshua, and Michel Siffre. "Caveman: An Interview with Michel Siffre." *Cabinet.* Summer 2008. https://www.cabinetmagazine.org/issues/30/foer_siffre.php.

Fuller, R. Buckminster. *Critical Path.* New York: St Martin's Press, 1981.

Garrett, Bradley. *Bunker: Building for the End Times.* New York: Scribner, 2020.

Hallman, J. C. "A House Is a Machine to Live In." *The Believer.* October 1, 2009. https://believermag.com/a-house-is-a-machine-to-live-in/.

Kean, Sam. "How Not to Deal with Murder in Space." *Slate.* July 15, 2020. https://slate.com/technology/2020/07/arctic-t3-murder-space.html.

National Technical Reports Library. *Triton City: A Prototype Floating Community*. U.S. Department of Housing and Urban Development. November 1968. https://ntrl.ntis.gov/NTRL/dashboard/searchResults/titleDetail/PB180051.xhtml.

Piccard, Bertrand. "Breitling Orbiter 3: GOSH Documentary." Accessed May 2021. YouTube video, 54:06. https://www.youtube.com/watch?v=kSmrHsG2v8I.

"Polar Explorer, Stabbed by a Colleague, Made Peace with Him in Court." *RAPSI News*. February 8, 2019. http://www.rapsinews.ru/judicial_news/20190208/294778187.html.

Poynter, Jane. *The Human Experiment: Two Years and Twenty Minutes Inside Biosphere 2*. New York: Basic Books, 2006.

seasteading. "THE FIRST SEASTEADERS 6: Fleeing The Death Threat." February 29, 2020. YouTube video, 10:59. https://www.youtube.com/watch?v=OovkeOuZsqU.

Silverstone, Sally. *Eating In: From the Field to the Kitchen in Biosphere 2*. Oracle, Arizona: Biosphere Foundation, 1993.

Stoll, Steven. *The Great Delusion: A Mad Inventor, Death in the Tropics, and the Utopian Origins of Economic Growth*. New York: Hill and Wang, 2008.

StrangerHopeful. "Can Cloud Nine Be Built?" *Stack Exchange*. December 30, 2016. https://worldbuilding.stackexchange.com/questions/36667/can-cloud-nine-be-built.

Tucker, Reed. "How Six Scientists Survived 'Living on Mars' for a Year." *New York Post*. November 14, 2020. https://nypost.com/2020/11/14/how-six-scientists-survived-living-on-mars-for-a-year.

Warner, Andy, and Sofie Louise Dam. *This Land Is My Land: A Graphic History of Big Dreams, Micronations, and Other Self-Made States*. San Francisco: Chronicle Books, 2019.

Wolf, Matt, dir. *Spaceship Earth*. 2020.

CHAPTER 2: HOW TO START YOUR OWN COUNTRY

Abdel-Motaal, Doaa. *Antarctica: The Battle for the Seventh Continent*. Santa Barbara, CA: Praeger, 2016.

Adewunmi, Bim. "I Claim This Piece of Africa for My Daughter, Princess Emily." *The Guardian*. July 15, 2014. https://www.theguardian.com/lifeandstyle/shortcuts/2014/jul/15/claim-piece-africa-for-daughter-princess-emily-sudan.

Arnold, Carrie. "In Splendid Isolation: The Research Voyages That Prepared Us for the Pandemic." *Nature*. May 14, 2020. https://www.nature.com/articles/d41586-020-01457-8.

Bueno de Mesquita, Bruce, and Alastair Smith. *The Dictator's Handbook: Why Bad Behavior Is Almost Always Good Politics*. New York: PublicAffairs, 2011.

Heuermann, Christoph. "Held Hostage by Bedouins in Terra Nullius: My Trek to and Escape from Bir Tawil." *Christoph Today*. December 8, 2019. https://christoph.today/sudan-bir-tawil.

Krakauer, Jon. *Into the Wild*. New York: Villard, 1996.

McHenry, Travis. *The Rise, Fall, and Rebirth of Westarctica*. Self-published, 2017. https://drive.google.com/file/d/1zYn3oMmFpQoSElMVd-ryMvYp3YKegIqq/view.

Secretariat of the Antarctic Treaty. "Key Documents of the Antarctic Treaty System." 2021. Accessed May 2021. https://www.ats.aq/e/key-documents.html.

———. "The Protocol on Environmental Protection to the Antarctic Treaty." October 4, 1991. Accessed May 2021. https://www.ats.aq/e/protocol.html.

Shenker, Jack. "Welcome to the Land That No Country Wants." *The Guardian*. March 3, 2016. https://www.theguardian.com/world/2016/mar/03/welcome-to-the-land-that-no-country-wants-bir-tawil.

Strauss, Erwin S. 1984. *How to Start Your Own Country: How You Can Profit from the Coming Decline of the Nation State*. Port Townsend, WA: Loompanics Unlimited, 1984.

CHAPTER 3: CLONING DINOSAURS, AND SOME PTERRIBLE NEWS FOR ALL WHO'D DARE OPPOSE YOU

Church, George M. *Regenesis: How Synthetic Biology Will Reinvent Nature and Ourselves*. New York: Basic Books, 2012.

Dell'Amore, Christine. "New 'Chicken from Hell' Dinosaur Discovered." *National Geographic*. March 19, 2014. https://www.nationalgeographic.com/news/2014/3/140319-dinosaurs-feathers-animals-science-new-species.

Department of Vertebrate Zoology, National Museum of Natural History. "The Passenger Pigeon." *Smithsonian*. March 2001. Accessed May 2021. https://www.si.edu/spotlight/passenger-pigeon.

Ewing, Jeff. "What If Jurassic World Were Real? 3 Hidden Economic Consequences." *Forbes*. June 22, 2018. https://www.forbes.com/sites/jeffewing/2018/06/22/3-hidden-economic-consequences-of-jurassic-world.

Horner, Jack, and James Gorman. *How to Build a Dinosaur*. New York: Dutton, 2009.

Kolbert, Elizabeth. *The Sixth Extinction: An Unnatural History*. New York: Henry Holt and Company, 2014.

Kornfeldt, Torill. *The Re-Origin of Species: A Second Chance for Extinct Animals*. Translated by Fiona Graham. Brunswick, Victoria, Australia: Scribe, 2018.

Yeoman, Barry. "Why the Passenger Pigeon Went Extinct." *Audobon*. May–June 2014. https://www.audubon.org/magazine/may-june-2014/why-passenger-pigeon-went-extinct.

CHAPTER 4: CONTROLLING THE WEATHER FOR THE PERFECT CRIME

Fleming, James Rodger. *Fixing the Sky: The Checkered History of Weather and Climate Control*. New York: Columbia University Press, 2010.

Goodell, Jeff. *How to Cool the Planet: Geoengineering and the Audacious Quest to Fix Earth's Climate*. New York: Houghton Mifflin Harcourt, 2010.

Morton, Oliver. *The Planet Remade: How Geoengineering Could Change the World*. Princeton, NJ: Princeton University Press, 2017.

Roser, Max, Esteban Ortiz-Ospina, and Hannah Ritchie. "Life Expectancy." Our World in Data. October 2019. Accessed May 2021. https://ourworldindata.org/life-expectancy.

Smith, Wake, and Gernot Wagner. "Stratospheric Aerosol Injection Tactics and Costs in the First 15 Years of Deployment." *Environmental Research Letters* 13, no. 12 (2018). https://doi.org/10.1088/1748-9326/aae98d.

Tamasy, Paul, and Aaron Mendelsohn. *Air Bud*. 1997. Directed by Charles Martin Smith. Produced by Robert Vince and William Vince. Starring Buddy the Golden Retreiver as "Air Bud."

United Nations. "Convention on the Prevention of Marine Pollution by Dumping of Wastes and Other Matter." International Maritime Organization. 1972. Accessed May 2021. https://www.imo.org/en/OurWork/Environment/Pages/London-Convention-Protocol.aspx.

———. "Convention on the Prohibition of Military or Any Other Hostile Use of Environmental Modification Techniques." United Nations Treaty Collection. December 1976. Accessed May 2021. https://treaties.un.org/Pages/ViewDetails.aspx?src=TREATY&mtdsg_no=XXVI-1&chapter=26&clang=_en.

———. "Universal Declaration of Human Rights." United Nations. December 1948. Accessed May 2021. https://www.un.org/en/about-us/universal-declaration-of-human-rights.

Victor, David G., M. Granger Morgan, Jay Apt, John Steinbruner, and Katharine Ricke. "The Geoengineering Option: A Last Resort Against Global Warming?" *Foreign Affairs* 88, no. 2 (March/April 2009): 64–76.

Wallace-Wells, David. *The Uninhabitable Earth: Life After Warming*. New York: Tim Duggan Books, 2019.

CHAPTER 5: SOLVING ALL YOUR PROBLEMS BY DRILLING TO THE CENTER OF THE PLANET TO HOLD THE EARTH'S CORE HOSTAGE

Kaplan, Sarah. "How Earth's Hellish Birth Deprived Us of Silver and Gold." *The Washington Post*. September 27, 2017. https://www.washingtonpost.com/news/speaking-of-science/wp/2017/09/27/how-earths-hellish-birth-deprived-us-of-silver-and-gold.

Köhler, Nicholas. "The Incredible True Story Behind the Toronto Mystery Tunnel." *Maclean's*. March 20, 2015. https://www.macleans.ca/society/elton-mcdonald-and-the-incredible-true-story-behind-the-toronto-mystery-tunnel.

Krugman, Paul. "Three Expensive Milliseconds." *The New York Times*. April 13, 2014. https://www.nytimes.com/2014/04/14/opinion/krugman-three-expensive-milliseconds.html.

Lewis, Michael. *Flash Boys: A Wall Street Revolt*. New York: W. W. Norton & Company, 2014.

Mining Technology. "Top 10 Deep Open-Pit Mines." September 26, 2013. https://www.mining-technology.com/features/feature-top-ten-deepest-open-pit-mines-world.

Osadchiy, A. "Kola Superdeep Borehole." *Science and Life*. November 5, 2002. https://www.nkj.ru/archive/articles/4172.

Piesing, Mark. "The Deepest Hole We Have Ever Dug." *BBC Future*. May 6, 2019. https://www.bbc.com/future/article/20190503-the-deepest-hole-we-have-ever-dug.

Warnica, Richard. "'There Is No Criminal Offense for Digging a Hole': Police Won't Speculate on Mystery Tunnel near Pan Am Site." *National Post*. February 24, 2015. https://nationalpost.com/news/toronto/there-is-no-criminal-offence-for-digging-a-hole-police-refuse-to-speculate-on-mystery-tunnel-near-pan-am-site.

World Gold Council. "How Much Gold Has Been Mined?" *About Gold*. 2020. https://www.gold.org/about-gold/gold-supply/gold-mining/how-much-gold.

CHAPTER 6: TIME TRAVEL

I promise I'll update this section just as soon as I figure out time travel, *if not sooner*.

CHAPTER 7: DESTROYING THE INTERNET TO SAVE US ALL

Ball, James. *The Tangled Web We Weave: Inside the Shadow System That Shapes the Internet*. New York: Melville House, 2020.

Hill, Kashmir. "Goodbye Big Five." *Gizmodo*. Accessed May 2021. https://gizmodo.com/c/goodbye-big-five.

Kolitz, Daniel. "What Would It Take to Shut Down the Entire Internet?" *Gizmodo*. September 30, 2019. https://gizmodo.com/what-would-it-take-to-shut-down-the-entire-internet-1837984019.

Peters, Jay. "Prolonged AWS Outage Takes Down a Big Chunk of the Internet." *The Verge*. November 25, 2020. https://www.theverge.com/2020/11/25/21719396/amazon-web-services-aws-outage-down-internet.

Sanchez, Julian. "Trump Is Looking for Fraud in All the Wrong Places." *The Atlantic*. December 12, 2020. https://www.theatlantic.com/ideas/archive/2020/12/trump-looking-fraud-all-wrong-places/617366.

Steinberg, Jospeh. "Massive Internet Security Vulnerability—Here's What You Need to Do." *Forbes*. April 10, 2014. https://www.forbes.com/sites/josephsteinberg/2014/04/10/massive-internet-security-vulnerability-you-are-at-risk-what-you-need-to-do.

Tardáguila, Cristina. "Electronic Ballots Are Effective, Fast and Used All Over the World—So Why Aren't They Used in the U.S.?" *Poynter*. November 4, 2020. https://www.poynter.org/fact-checking/2020/electronic-ballots-are-effective-fast-and-used-all-over-the-world-so-why-arent-used-in-the-u-s.

Zerodium. "Our Exploit Acquisition Program." *Zerodium*. Accessed May 2021. https://zerodium.com/program.html.

Zetter, Kim. *Countdown to Zero Day: Stuxnet and the Launch of the World's First Digital Weapon*. New York: Crown, 2014.

CHAPTER 8: HOW TO BECOME IMMORTAL AND LITERALLY LIVE FOREVER

Achenbaum, W. Andrew. "America as an Aging Society: Myths and Images." *Daedalus* 115, no. 1 (Winter 1986): 13–30. https://www.jstor.org/stable/20025023.

Bacon, Francis. *The Historie of Life and Death, With Observations Naturall and Experimentall for the Prolonging of Life*. London: I. Okes, for Humphrey Mosley, 1638.

Begley, Sharon. "After Ghoulish Allegations, a Brain-Preservation Company Seeks Redemption." *Stat*. January 30, 2019. https://www.statnews.com/2019/01/30/nectome-brain-preservation-redemption.

Bernstein, Anya. *The Future of Immortality: Remaking Life and Death in Contemporary Russia*. Princeton, NJ: Princeton University Press, 2019.

Boyle, Robert. *Some Considerations Touching the Usefulnesse of Experimental Naturall Philosophy, Propos'd in Familiar Discourses to a Friend, by Way of Invitation to the Study of It*. Oxford: Oxford University, 1663.

———. "Tryals Proposed by Mr. Boyle to Dr. Lower, to Be Made by Him, for the Improvement of Transfusing Blood out of One Live Animal into Another." *Philosophical Transactions (1665–1678)* 1 (1666): 385–88. http://www.jstor.org/stable/101547.

Chrisafis, Angelique. "Freezer Failure Ends Couple's Hopes of Life After Death." *The Guardian*. March 16, 2006. https://www.theguardian.com/science/2006/mar/17/france.internationalnews.

Darwin, Mike. "Evaluation of the Condition of Dr. James H. Bedford After 24 Years of Cryonic Suspension." *Cryonics*. August 1991. https://www.alcor.org/library/bedford-condition.

de Longeville, Harcouet. *Long Livers: A Curious History of Such Persons of Both Sexes who have liv'd several Ages, and grown Young again: With the rare Secret of Rejuvenescency of Arnoldus de Villa Nova, and a great many approv'd and invaluable Rules to prolong Life: as also How to prepare the Universal Medicine*. London: J. Holland, 1772.

Friedman, David M. *The Immortalists: Charles Lindbergh, Dr. Alexis Carrel, and Their Daring Quest to Live Forever*. New York: Ecco, 2008.

Haycock, David Boyd. *Mortal Coil: A Short History of Living Longer*. New Haven, CT: Yale University Press, 2009.

Henderson, Felicity. "What Scientists Want: Robert Boyle's To-Do List." *The Royal Society*. August 2010. https://royalsociety.org/blog/2010/08/what-scientists-want-boyle-list.

KrioRus. "Animals—Cryopatients." *KrioRus*. Accessed May 2021. https://kriorus.ru/Zhivotnye-kriopacienty.

———. "List of People Cryopreserved in KrioRus." *KrioRus*. Accessed May 2021. https://kriorus.ru/Krionirovannye-lyudi.

Maxwell-Stuart, P. G. *The Chemical Choir: A History of Alchemy*. London: Bloomsbury Academic, 2012.

McIntyre, Robert. "The Case for Glutaraldehyde: Structural Encoding and Preservation of Long-Term Memories." *Nectome*. Accessed May 2021. https://nectome.com/the-case-for-glutaraldehyde-structural-encoding-and-preservation-of-long-term-memories.

Perry, R. Michael. "Suspension Failures: Lessons from the Early Years." *Cryonics*. February 1992. https://www.alcor.org/library/suspension-failures-lessons-from-the-early-years.

Pontin, Jason. "Is Defeating Aging Only a Dream?" *Technology Review. The SENS Challenge*. July 11, 2006. http://www2.technologyreview.com/sens.

Redwood, Zander. "Living to 1000: An Interview with Aubrey de Grey." *80000 Hours*. April 12, 2012. https://80000hours.org/2012/04/living-to-1000-an-interview-with-aubrey-de-grey.

Roser, Max, Esteban Ortiz-Ospina, and Hannah Ritchie. "Life Expectancy." Our World in Data. October 2019. Accessed May 2021. https://ourworldindata.org/life-expectancy.

SENS Research Foundation. "An Introduction to SENS Research." Accessed May 2021. https://www.sens.org/our-research.

Shaw, Sam. "You're As Cold As Ice." *This American Life.* April 18, 2008. https://www.thisamericanlife.org/354/mistakes-were-made.

Standiford, Les. *Meet You in Hell: Andrew Carnegie, Henry Clay Frick, and the Bitter Partnership That Changed America.* New York: Crown, 2006.

Svyatogor, Alexander. "Biocosmist Poetics." In *Russian Cosmism,* edited by Boris Groys. Cambridge, MA: MIT Press, 2018.

Walter, Chip. *Immortality, Inc.: Renegade Science, Silicon Valley Billions, and the Quest to Live Forever.* Washington, D.C.: National Geographic, 2020.

Weiner, Jonathan. *Long for This World: The Strange Science of Immortality.* New York: Ecco, 2011.

CHAPTER 9: ENSURING YOU ARE NEVER, EVER, *EVER* FORGOTTEN

Benford, Gregory. *Deep Time.* New York: HarperPerennial, 2000.

Blakeslee, Sandra. "Lost on Earth: Wealth of Data Found in Space." *The New York Times.* March 20, 1990. https://www.nytimes.com/1990/03/20/science/lost-on-earth-wealth-of-data-found-in-space.html.

Brannen, Peter. *The Ends of the World: Volcanic Apocalypses, Lethal Oceans, and Our Quest to Understand Earth's Past Mass Extinctions.* New York: Ecco, 2017.

Brook, Pete. "In Billions of Years, Aliens Will Find These Photos in a Dead Satellite." *Wired.* October 30, 2012. https://www.wired.com/2012/10/the-last-pictures.

Bump, Philip. "How to Put Trump on Mount Rushmore, Something He's Never Even Thought About." *The Washington Post.* July 26, 2017. https://www.washingtonpost.com/news/politics/wp/2017/07/26/how-to-put-trump-on-mount-rushmore-something-hes-never-even-thought-about.

Chiba, Sanae, et al. "Human Footprint in the Abyss: 30 Year Records of Deep-Sea Plastic Debris." *Marine Policy* 96 (2018): 204–8. https://doi.org/10.1016/j.marpol.2018.03.022.

Culp, Justin. "Archeological Inventory at Tranquility Base." *Lunar Legacy Project.* April 8, 2002. Accessed May 2021. http://spacegrant.nmsu.edu/lunarlegacies/artifactlist.html.

Diaz, Jesus. "All the American Flags On the Moon Are Now White." *Gizmodo.* July 31, 2012. https://gizmodo.com/all-the-american-flags-on-the-moon-are-now-white-5930450.

Drogin, Marc. *Anathema!: Medieval Scribes and the History of Book Curses.* Montclair, NJ: Abner Schram, 1983.

Gilbert, Samuel. "The Man Who Helped Design a 10,000-Year Nuclear Waste Site Marker." *Vice.* April 26, 2018. https://www.vice.com/en/article/9kgjze/jon-lomberg-nuclear-waste-marker-v25n1.

Gilbertson, Scott. "The Very First Website Returns to the Web." *Wired.* April 30, 2013. https://www.wired.com/2013/04/the-very-first-website-returns-to-the-web.

Haskoor, Michael. "A Space Jam, Literally: Meet the Creative Director Behind NASA's 'Golden Record,' an Interstellar Mixtape." *Vice.* April 4, 2015. https://www.vice.com/en/article/rgpj5j/the-golden-record-ann-druyan-interview.

Hera, Stephen C., Detlof von Winterfeldt, and Kathleen M. Trauth. "Expert Judgment on Inadvertent Human Intrusion into the Waste Isolation Pilot Plant." Sandia National Laboratories Report SAND90-3063. December 1991.

Holman, E. W., S. Wichmann, C. H. Brown, V. Velupillai, A. Müller, and D. Bakker. "Explorations in Automated Language Classification." *Folia Linguistica* 42 (2008): 3–4. doi:10.1515/flin.2008.331.

Jet Propulsion Laboratory. "The Golden Record." Accessed May 2021. https://voyager.jpl.nasa.gov/golden-record.

Lomberg, Jon. 2007. "A Portrait of Humanity." 2007. Accessed May 2021. https://www.jonlomberg.com/articles/a_portrait_of_humanity.html.

Marchant, Jo. "In Search of Lost Time." *Nature* 444 (2006): 534–38. https://doi.org/10.1038/444534a.

Memorial Museum of Cosmonautics. "Lunar Pennants of the USSR." January 29, 2016. Accessed May 2021. https://kosmo-museum.ru/news/lunnye-vympely-sssr.

NASA News. "Project Lageos." Press Kit, NASA. 1976. https://lageos.gsfc.nasa.gov/docs/1976/NASA_LAGEOS_presskit_e000045273.pdf.

Nurkiyazova, Sevindj. "The English Word That Hasn't Changed in Sound or Meaning in 8,000 Years." *Nautilus.* May 13, 2019. http://nautil.us/blog/the-english-word-that-hasnt-changed-in-sound-or-meaning-in-8000-years.

Paglen, Trevor. *The Last Pictures.* Berkeley and Los Angeles: University of California Press, 2012.

Parkinson, R. B. *The Rosetta Stone.* London: British Museum Press, 2005.

Radioactive Waste Management Committee. "Preservation of Records, Knowledge and Memory across Generations: A Literature Survey on Markers and Memory Preservation for Deep Geological Repositories, Swiss Nuclear Energy Agency." Nuclear Energy Agency. 2013. https://www.oecd-nea.org/jcms/pl_19357.

Safire, Bill. "In Event Of Moon Disaster." July 18, 1969. https://www.archives.gov/files/presidential-libraries/events/centennials/nixon/images/exhibit/rn100-6-1-2.pdf.

Sagan, Carl. *Cosmos.* New York: Random House, 1980.

Sebeok, Thomas A. "Communication Measures to Bridge Ten Millennia." Technical Report, Research Center for Language and Semiotic Studies. 1984. https://doi.org/10.2172/6705990.

Taliaferro, John. *Great White Fathers: The True Story of Gutzon Borglum and His Obsessive Quest to Create the Mt. Rushmore National Monument.* New York: PublicAffairs, 2004.

The Museum of Modern Art. "The Moon Museum: Various Artists with Andy Warhol, Claes Oldenburg, David Novros, Forrest Myers, Robert Rauschenberg, John Chamberlain." *The MoMA Collection.* Accessed May 2021. https://www.moma.org/collection/works/62272.

Trauth, Kathleen M., Stephen C. Hora, and Robert V. Guzowski. "Expert Judgment on Markers to Deter Inadvertent Human Intrusion into the Waste Isolation Pilot Plant." Sandia National Laboratories Report SAND92-1382. 1993. http://doi.org/10.2172/10117359.

Trosper, Jaime. "Death & Decomposition in Space." *Futurism.* December 18, 2013. https://futurism.com/death-decomposition-in-space.

Ward, Peter D., and Donald Brownlee. *The Life and Death of Planet Earth: How the New Science of Astrobiology Charts the Ultimate Fate of Our World.* New York: Holt Paperbacks, 2004.

Weisman, Alan. *The World Without Us.* New York: St. Martin's Thomas Dunne Books, 2007.

INDEX